GRID INTEGRATION OF ELECTRIC VEHICLES IN OPEN ELECTRICITY MARKETS

GRID INTEGRATION OF ELECTRIC VEHICLES IN OPEN ELECTRICITY MARKETS

Qiuwei Wu
Technical University of Denmark, Lyngby

WILEY

Library of Congress Cataloging-in-Publication Data

Grid integration of electric vehicles in open electricity markets / edited by Professor Qiuwei Wu.
    pages cm
  Includes bibliographical references and index.
  ISBN 978-1-118-44607-2 (cloth)
 1. Electric automobiles–Power supply–Forecasting.   2. Electric automobiles–Power supply–Economic aspects.   3. Electric power plants–Load.   4. Electric power consumption.   5. Electric utilities.   I. Wu, Qiuwei.
  TL220.G75 2013
  333.793′2–dc23

                                                                                    2012050750

A catalogue record for this book is available from the British Library

ISBN: 978-1-118-44607-2

Typeset in 10/12pt Times by Aptara Inc., New Delhi, India

Printed and bound in Singapore by Markono Print Media Pte Ltd

1   2013

Contents

List of Contributors

Jonas Åkerman, Royal Institute of Technology (KTH), Division of Environmental Strategies Research, Stockholm, Sweden

Martin Albrecht, Royal Institute of Technology (KTH), Division of Environmental Strategies Research, Stockholm, Sweden

Birgitte Bak-Jensen, Department of Energy Technology, Aalborg University, Aalborg, Denmark

Christian Bang, Ea Energy Analyses A/S, Copenhagen, Denmark

Andreas Bjerre, DONG Energy Sales and Distribution, Virum, Denmark

Trine Krogh Boomsma, Department of Mathematical Sciences, University of Copenhagen, Copenhagen, Denmark

Stefanos Delikaraoglou, Technical University of Denmark, Lyngby, Denmark

Yi Ding, Centre for Electric Power and Energy (CEE), Department of Electrical Engineering, Technical University of Denmark, Lyngby, Denmark

Lars Henrik Hansen, DONG Energy, Power Concept Optimization, Gentofte, Denmark

Camilla Hay, Ea Energy Analyses A/S, Copenhagen, Denmark

Jakob Munch Jensen, DONG Energy, Power Concept Optimization, Gentofte, Denmark

Søren Højgaard Jensen, Department of Energy Conversion and Storage, Technical University of Denmark, Roskilde, Denmark

Nina Juul, DTU Management Engineering, Technical University of Denmark, Roskilde, Denmark

Arne Hejde Nielsen, Centre for Electric Power and Energy (CEE), Department of Electrical Engineering, Technical University of Denmark, Lyngby, Denmark

Måns Nilsson, Royal Institute of Technology (KTH), Division of Environmental Strategies Research, Stockholm, Sweden; Stockholm Environment Institute (SEI), Stockholm, Sweden

Niamh O'Connell, Centre for Electric Technology, Department of Electrical Engineering, Technical University of Denmark, Copenhagen, Denmark

Jacob Østergaard, Centre for Electric Power and Energy (CEE), Department of Electrical Engineering, Technical University of Denmark, Lyngby, Denmark

Jayakrishnan R. Pillai, Department of Energy Technology, Aalborg University, Aalborg, Denmark

Claus Nygaard Rasmussen, Centre for Electric Technology, Department of Electrical Engineering, Technical University of Denmark, Lyngby, Denmark

Charlotte Søndergren, Danish Energy Association, Lyngby, Denmark

Mikael Togeby, Ea Energy Analyses A/S, Copenhagen, Denmark

Qiuwei Wu, Centre for Electric Power and Energy (CEE), Department of Electrical Engineering, Technical University of Denmark, Lyngby, Denmark

Guang Ya Yang, Centre for Electric Technology, Department of Electrical Engineering, Technical University of Denmark, Lyngby, Denmark

Preface

Environmental concerns and the quest for energy supply independence have resulted in increasing penetration of renewable energy sources (RES) and a move toward electrification of transportation. Consequently, electric vehicles (EVs) are expected to play a significant role in the future power systems and impact distribution networks. Increased use of EVs will reduce greenhouse gas emissions from the transport sector by replacing conventional internal combustion engine vehicles while also serving as distributed energy storage that can mitigate uncertainties arising from intermittent RES.

This book deals with the grid integration issues of EVs into the open electricity market in order to efficiently realize the large-scale deployment of EVs in the future power systems. The book starts with the introduction of policy drivers and environmental impacts of the electrification of vehicles. It is followed by the EV integration options in the current Nordic electricity market and the possible electricity market development in order to facilitate EV integration. The investment and operation of an integrated electricity and transport system is discussed in detail as well. Following that, the EV charging management using dynamic programming and EV fleet operator concept are described. In the end, the benefits of using EVs to provide regulating power, frequency-control reserves, and voltage support are presented along with the impact of EVs integration on the operation of electric distribution networks and battery degradation, and congestion management within electric distribution networks.

First of all, I would like to thank Peter Mitchell from John Wiley & Sons Ltd for initializing the idea of writing a book on the EV grid integration issues and managing the whole process. I also would like to thank a number of employees from John Wiley & Sons Ltd, namely Assistant Editor Laura Bell, Project Editor Liz Wingett, Senior Project Editor Richard Davies, and Assistant Production Editor Nur Wahidah Binte Abdul Wahid, for their endless support. I would like to thank the Edison project consortium for their permission to use the project results to contribute Chapters 2, 3, 6, 7, 9, 10, and 11.

Last but not least, I would like to thank all the book chapter contributors.

Qiuwei Wu
Copenhagen, Denmark

List of Abbreviations

ACE	area control error
AGC	automatic generation control
BEV	battery electric vehicle
BMS	battery management system
BRP	balance responsible party
CC	constant current
CCGT	combined cycle gas turbine
CHP	combined heat and power
COSOC	contractural state of charge
CPP	critical peak price
DAM	day-ahead market
DOD	depth of discharge
DSO	distribution system operator
DT	dynamic tariff
EC	equivalent circuit
EDV	electric-drive vehicle
EIS	electrochemical impedance spectroscopy
EMO	electric mobility operator
EV	electric vehicle
FCEV	fuel-cell plug-in hybrid electric vehicle
FCV	fuel-cell vehicle
FO	fleet operator
GHG	greenhouse gas
GIV	grid-integrated vehicle
HEV	hybrid electric vehicle
ICE	internal combustion engine
LBR	load balance responsible
LFC	load frequency control
LFP	lithium iron phosphate
LMP	locational marginal pricing
LV	low voltage
MV	medium voltage
NEDC	New European Drive Cycle
NMC	nickel–manganese–cobalt oxide

NOIS	Nordic Operational Information System
OCV	open-circuit voltage
OEM	original equipment manufacturer
OPF	optimal power flow
PBR	production balance responsible
PHEV	plug-in hybrid electric vehicle
RES	renewable energy sources
REV	range-extended electric vehicle
RSC	resources, scheduling and commitment
RTM	real-time market
RTO	regional transmission organization
SC	short circuit
SEA	Swedish Energy Agency
SLD	single line diagram
SME	small and medium enterprise
SOC	state of charge
SOH	state of health
TDM	time-domain method
TIS	technological innovation system
TOU	time-of-use
TSO	transmission system operator
V2G	vehicle from the grid
V2G	vehicle-to-grid
VAT	value-added tax
VPP	virtual power plant
WTP	willingness-to-pay

1

Electrification of Vehicles: Policy Drivers and Impacts in Two Scenarios

Martin Albrecht,[1] Måns Nilsson[1,2] and Jonas Åkerman[1]

[1] *Royal Institute of Technology (KTH), Division of Environmental Strategies Research, Stockholm, Sweden*
[2] *Stockholm Environment Institute (SEI), Stockholm, Sweden*

1.1 Introduction

Over the last 10 years, the interest for low-carbon vehicle technologies has surged among both governments and automotive manufacturers across and beyond the European Union (EU). Great hopes have been put, first, on biofuel vehicles and more recently (as the enthusiasm for biofuels cooled off) on electric vehicles (EVs) and hybrid electric vehicles (HEVs) as key technologies to mitigate climate change, enhance energy security and nurture new industry branches within the automotive sector. In particular, in the Nordic region, where electricity production has a relatively minor fossil input on average, electrification of transport has been seen as a key strategy to reduce CO_2 emissions from the transport sector.

However, while the market penetration for biofuel vehicles has been relatively high in some countries, the corresponding increases in electrification of vehicles have not materialized so far. An important reason for this is that vehicle prices remain considerably higher for EVs and HEVs compared with internal combustion engine (ICE)-based vehicles mostly due to high lithium-ion battery prices. Also, the shape of the learning curve and associated future costs remain uncertain and predictions vary strongly [1–3]. Lack of experience with battery durability under different climatic and driving conditions poses a significant risk for early adopters investing in a new EV car. Additionally, battery electric vehicles (BEVs), and in some cases also plug-in hybrid electric vehicle (PHEVs) or range-extended electric vehicles

Grid Integration of Electric Vehicles in Open Electricity Markets, First Edition. Edited by Qiuwei Wu.
© 2013 John Wiley & Sons, Ltd. Published 2013 by John Wiley & Sons, Ltd.

(REVs), require new infrastructure (for charging and to some extent for the upgrade of the local power grid) and different driving behaviour. As a result, there are major uncertainties in (a) future forecasts about BEV/PHEV/REV market penetration, (b) what policy frameworks are needed to facilitate the market uptake of these vehicles and (c) what are ultimately the climate implications of these forecasts. We do know that, over the coming years, BEV/PHEV technology will require public governance measures of different types, both to induce innovation and market uptake, and to control and mitigate possible environmental and social consequences.

This chapter addresses these uncertainties in the context of the Nordic region (Denmark, Finland, Norway and Sweden), through focusing our discussion on the following questions:

- How do policies, goals and targets within and across the Nordic countries compare against industry, government and expert forecasts about market uptake?
- What policy or broader governance initiatives are likely needed to have a plausible chance of reaching a breakthrough scenario?
- What are the climate impacts of our scenarios and what are the implications for the attainment of climate targets?

The chapter unfolds as follows. In Section 1.2 we present a review of policies and key targets in the Nordic countries and the EU, and discuss to what extent they align with or deviate from industry and expert estimates of how the systems can grow. On the basis of this, Section 1.3 elaborates scenarios of EV development in the EU and analyses the energy and climate impacts of the two scenarios, given different assumptions relating to power supply in the Nordic region. Section 1.4 examines what policy drivers might be needed to enable the breakthrough scenarios, using a technological innovation system (TIS) perspective to describe the processes, drivers and developments needed in policy and technology. Section 1.5 summarizes our results and conclusions.

1.2 Policy Drivers, Policies and Targets

Across the EU and globally, policy makers' interests in the electrification of vehicles have surged. Most EU countries have presented national development plans and targets for EVs. The interest is related to at least three political priorities.

The first concerns *climate change mitigation*. In the Nordic countries, total passenger car emissions in 2010 accounted for 14.10% of total emissions in Denmark, 11.15% in Finland, 12.31% in Norway and 23.05% in Sweden (see Figure 1.1) [4–9]. It is worth noting that this makes Sweden the second worst in the EU27 when just looking at the percentage. This is partly a result of Sweden having relatively lower emissions percentages in other sectors. However, it still indicates that it is especially in this sector that Sweden still has much to gain from mitigation measures.

The emission share of passenger cars within road transport is decreasing in most Nordic countries, while emissions from light and heavy trucks are increasing (see also Figure 1.2) [4–9]. The numbers are, however, overshadowed by the financial and economic crisis which reduced economic activity in the other road transport modes. Overall, the long-term trend indicates that some of the transport work is shifted between road transport modes but also that the environmental performance of passenger cars is improving more quickly.

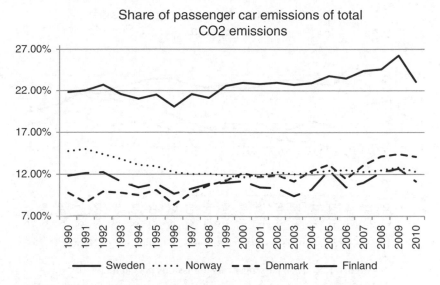

Figure 1.1 Passenger car's share of total emissions in the Nordic countries [4–9].

In absolute terms, CO_2 emissions from passenger cars stayed on a relatively high but stable level for Sweden and there are signs of a downward trend. The other Nordic countries are still growing in absolute CO_2 emissions from passenger cars, although from a much lower base. If one looks at the passenger car emissions per capita numbers, the Swedish downward trend becomes more obvious [4–10]. Norway has been able to stabilize its emissions, while Denmark has almost succeeded in doing so. Before the financial and economic crisis, Finland was on a clear upward trend (see also Figure 1.3).

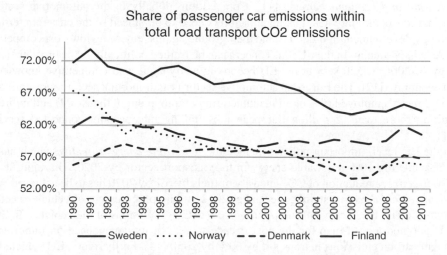

Figure 1.2 Passenger car's share of total road transport emissions in the Nordic countries [4–9].

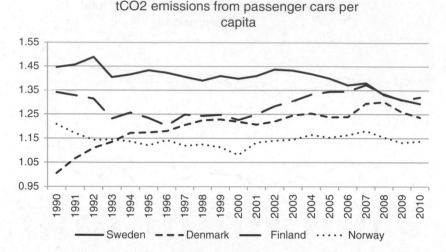

Figure 1.3 Total passenger cars' emissions per capita [4–10].

Generally, rapid action is required to reduce passenger car emissions in line with ratified climate change goals. Otherwise, extrapolating the current function of environmental performance of the average passenger car in the fleet, we will not see a carbon-neutral road transport sector within the next couple of decades. The data also suggest that, even though passenger cars are the most important challenge right now, we will also have to tackle light and heavy trucks in the near future if one wants to counter given growth trends (see Figure 1.2).

The second political priority concerns *energy security*. Overall, transport accounts for around one-third of energy consumption, and with its heavy reliance on fossil fuels the sector is vulnerable to oil supply and connected price changes. The electrification of vehicles is a prime strategy to decrease the reliance on imported fossil fuels. The third concerns *innovation, job creation and economic growth* [11]. Competition globally in the automotive sector is fierce, and it is commonly held that manufacturers need to be 'ahead of the curve' in terms of technology development in order to stand their ground against emerging low-cost competition from Asia in particular. In the EU, this concern can be framed politically in the broader Lisbon strategy of 2006, which set out the EU becoming a 'dynamic and competitive knowledge-based economy' [12]. The European automotive sector is an important sector representing 2.3 million directly employed (7% of all manufacturing employment in the EU27) and indirectly supporting more than 12 million European jobs (taking into account connected services, etc.) [13].

On the EU level, important policies include the *renewable energy directive* which has the goal of achieving 10% renewable energy in the transport sector by 2020. Through the *fuel quality directive*, a reduction of CO_2 intensity of fuels by 6% by 2020 has to be achieved. With the *clean vehicle directive* starting December 2012, public procurement of vehicles needs to take into account the energy consumption as well as CO_2 emissions of the vehicles. In 2011, the EU adopted a road-map for the next decade to reduce its dependence on imported oil and to cut carbon emissions in transport by 60% by 2050 [14]. Furthermore, EU vehicle CO_2 emissions regulations stipulate that 130 g/km (phased in, starting 2012) has to be met by 2015

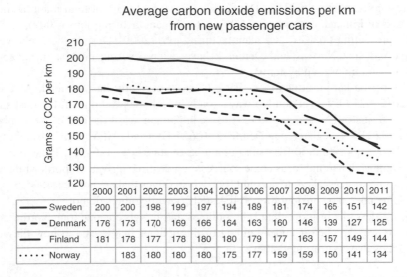

Figure 1.4   Average CO_2 emissions per kilometre from new passenger cars [19–21].

	2000	2001	2002	2003	2004	2005	2006	2007	2008	2009	2010	2011
Sweden	200	200	198	199	197	194	189	181	174	165	151	142
Denmark	176	173	170	169	166	164	163	160	146	139	127	125
Finland	181	178	177	178	180	180	179	177	163	157	149	144
Norway		183	180	180	180	175	177	159	159	150	141	134

and that 95 g/km very likely has to be fulfilled by 2020 [15–17]. Furthermore, the European parliament has mentioned the possibility of setting a 75 g/km CO_2 target for 2025 [18]. To set those numbers into context, the current grams of CO_2 per kilometre data for the average new passenger cars in the Nordic countries can be seen in Figure 1.4 [19–21]. The graph shows that Sweden and Finland are clearly lagging behind Norway and Denmark. In fact, Denmark already is below the 2015 EU emission target.

Globally, as well as in the EU, the economic crisis since 2008 pressed for stimulus spending in the automotive sector. Governments have provided subsidies, loans and research and development (R&D) support, the latter typically oriented towards environmentally friendly cars. Piloting and demonstration projects have often been implemented in cooperation with the private sector and in cooperation between universities, public institutions, the power industry and the automotive industry both on the national level and the European level.

Tax incentives, such as CO_2-differentiated vehicle taxes and car rebates, have been introduced in many countries in the EU. However, the tax level can be very different from country to country, taking into account the full set of measures. Kley et al. found that, as of 2010, the EU countries could be grouped into three categories with respect to the total incentives provided when it comes to mid-sized cars [22, 23]:

- the leaders (incentive from €10 000 to €28 000: Denmark, Norway and Belgium);
- the followers (incentive from €4000 to €9000: Netherlands, Spain, UK, France, Switzerland and Austria);
- the laggards (with amounts ±€3000: Ireland, Greece, Italy, Germany, Sweden, Poland and Finland).

Among the Nordic countries, only Sweden has a significant automotive industry [24]. The sector directly employs roughly 72 000 people in Sweden, representing 10.7% of total manufacturing jobs (2009), 6331 in Denmark, representing 1.6% of total manufacturing jobs (2008), 7509 in Finland, representing 1.9% of total manufacturing jobs (2009), and 3300 in Norway, representing 1.4% of total manufacturing jobs (2009). Despite their relatively small automotive industry, Norway and Denmark have taken a strong interest in advancing EV technologies and innovation systems.

In terms of market introduction of EVs, Norway currently has the lead. At the end of October 2012, 9212 EVs were on Norway's roads, which makes it one of the most successful countries in terms of EVs per capita [25]. By comparison, as of the end of September 2012, there were 1320 BEVs registered in Denmark, 1067 BEVs and PHEVs as of the end of October 2012 in Sweden and about 60 BEVs in Finland as of June 2012 [26–31]. These numbers are, however, somewhat unreliable, as some sources include direct private imports while others do not. Also, some sources take into account four-wheel drives that are not classified as passenger cars and some take into consideration PHEVs/REVs while others do not.

Below, we describe in more detail the policies and targets for our four Nordic countries. Through this we get a better understanding of the policies that exist and how they compare with the policy drivers presented above.

1.2.1   Finland

Goals

Finland has so far not established a specific national goal for the introduction of EVs. However, the government has presented a climate and energy strategy where two goals are to reduce greenhouse gas (GHG) emissions from traffic and transport by 15% and to increase the energy efficiency of the transport sector by 9% from 2005 to 2020 [32]. The government has also developed a vision for 2050 in which the direct specific emissions of cars are supposed to reach 80–90 g/km CO_2 by 2030, 50–60 g/km CO_2 by 2040 and 20–30 g/km CO_2 by 2050 [33].

Policy Instruments

A vehicles tax reform began in 2008 which eventually is supposed to give consumers more choice on the level of tax when they buy new or used cars [34, 35]. Today, the registration tax and the annual vehicle tax are based on CO_2 emissions. The new registration tax was introduced in 2008 and the new annual vehicle tax in 2010 [36]. In 2012, the lowest registration tax level, for cars with 0 g/km CO_2, was reduced from 12.2% to 5% [34, 35, 37, 38]. The highest tax level was raised from 48.8% to 50%. Overall, the message is that cars with less than 110 g/km CO_2 will get a lower registration tax compared with the tax regime before. For new BEVs, this means that the previous registration tax is being reduced from €3660 to €1500 for a BEV that costs €30 000. The base tax within the annual vehicle tax is also based on CO_2 emissions and after the 1 April 2012 can vary between €43 and €606 per year [34].

The Finnish government has also identified the EV as a Finnish export opportunity [39]. Subsequently, in 2011, TEKES (the Finnish Funding Agency for Technology and Innovation) introduced a 5-year program for the development of concepts for the EV and connected infrastructure [38, 40]. The programme is called EVE – Electric Vehicle Systems programme – and also hopes to create a strong community around EVs in Finland [41]. The largest project in the portfolio is the Electric Traffic Helsinki Test Bed project, which among other targets has the aim to establish around 850 charging spots in the capital region and enable the driving of 400 EVs during a period of 4 years [42–44]. Other significant projects include EVELINA (National Test Environment for Electric Vehicles) [45], Eco Urban Living [46], SIMBe (Smart Infrastructures for Electric Mobility in Built Environments, which started in January 2010 and is funded by TEKES Sustainable community programme) [47], and the battery research programme SINi [48].

Industry Position

Finland has a major and experienced EV manufacturing facility through the company Valmet Automotive who mainly builds EVs for other brands; for example, the REV sports car Fisker Karma [49]. Furthermore, before its recent bankruptcy, the Think car was produced in Finland at the same factory [50]. Another Finnish EV manufacturer is the company AMC Motors with their model Sanifer [51]. Finland is also home to a larger battery manufacturer called European Batteries [52]. Fortum as the major Finnish utility is part of several pilot projects across the Nordic countries and is foremost driving developments in the smart and fast charging area [53, 54].

1.2.2 Sweden

Goals

The Swedish government has established the vision of a "fossil fuel independent" transport sector by 2030, but has no target for PHEV/BEV penetration. Industry groups have put forward a vision for 600 000 PHEVs and BEVs on Swedish roads by 2020 [55–57]. The 2030 government vision is currently not backed up by concrete road-maps, even though the government recently decided to develop such a road-map [56]. At the same time, different industry organizations have established scenarios [58, 59]. There is significant scepticism and uncertainty about those targets, and even government officials think that only a modest 20 000–85 000 PHEVs and BEVs by 2020 is actually achievable under current institutional conditions [55, 60, 61].

Policy Instruments

Sweden has implemented a number of separate policy measures that are targeted at environmental friendly cars in a seemingly technology neutral way. A major part of Sweden's policy package, and the debate around it, centres on the green car definition. Confusingly, different definitions persist, emanating from different institutional homes: the road transport law, the income tax law and from several municipalities developing their own definitions [62]. The road transport law primarily eliminates the yearly vehicle tax for private persons and professional organizations for a period of 5 years for all green cars introduced after 1 June 2009 (currently, the green car definition translates into 120 g/km CO_2 – or cars driven by alternative fuels with fuel consumption per 100 km of 9.2 L gasoline equivalents, 9.2 m^3 of gas or 37 kWh electricity). A new green car definition is scheduled to be implemented early in 2013.

For the income year 2012 and 2013 the income tax law foresees that the tax on the private benefit stemming from an employee-driven but company-owned BEV, PHEV or biogas car to be 40% less than a comparable average model. The reduction takes place after the tax level has already been reduced to the average model; but all in all, the total reduction cannot be higher than 16 000 SEK [63]. Ethanol cars, HEVs and a variety of other biofuels are only reduced to the tax level of a comparable average model but are not reduced further. In 2012, the government introduced a new 40 000 SEK subsidy for the purchase of 'super green' cars (less than 50 g/km CO_2). The budget will be sufficient to support the equivalent of about 5000 EVs [64, 65]. At the end of September 2012 the maximum budget for 2012, which was 20 million SEK, had been reached [66].

Additionally, Swedish government efforts are connected to research funding usually for larger industry players (e.g. Volvo, Saab) as well as several pilot projects across Sweden (e.g. Malmö, Gothenburg, Stockholm, Östersund, Sundsvall, Helsingborg) [67–72]. Those measures are co-financed with a 25–50% stake by the Strategic Vehicle Research and Innovation programme (FFI – a Vinnova-funded research programme) or the Swedish Energy Agency (SEA) [61]. Other significant incentives include the national procurement plan initiated by the city of Stockholm and Vattenfall and partly financed by SEA [73]. The purpose of the procurement is to allow the coordinated procurement of 6000 EVs for companies and public agencies.

Regulatory changes are made to enable EV introductions. Since February 2011, municipalities can reserve parking spots in public spaces for EVs [61, 74]. However, it is not allowed to discriminate different types of vehicles when it comes to parking fees [71]. As a way to accelerate charging infrastructure deployment, there is no longer a need to pay grid concession fees to the local grid company for connecting outside charging infrastructure (e.g. in malls) [75–79].

Industry Position

In Sweden, industry is primarily concerned with R&D of electric powertrains or aspects related to them. However, Volvo is on the verge of commercializing two cars, namely a BEV and a PHEV, the latter co-financed by Vattenfall. Similar to Volvo, Saab has also developed a BEV, but the future of this project due to the company's recent bankruptcy remains uncertain. The new owner expects that they will sell a Saab EV by 2014 [80]. The company EV Adapt is converting conventional cars to BEVs and there is also a company called Hybricon that will be selling electric buses. Otherwise, there are also a number of companies that are active in the charging infrastructure business (e.g. Park&Charge, ChargeStorm, Easycharge). Moreover, Sweden has, and has had, a number of demonstration programmes in which, for example, utilities have been major partners [53, 81].

1.2.3   Denmark

Goals

In 2009, the Danish parliament agreed on a common policy for a greener transport system [82]. The new Danish government recently adopted the goal to phase out all of the country's oil, coal and natural gas until 2050 and to provide 50% of the country's electricity by wind energy already by 2020 [83, 84].

Policy Instruments

The major EV instrument is the relief from registration fees until 2015 [85–87]. The registration fee on passenger cars in Denmark in 2011 is 105% of the value until 79 000 DKK and 180% of the value above [88], making such a tax relief a very strong incentive. Also, the annual taxation of cars has been reformed: the tax was previously calculated on the basis of car weight, but is now based on fuel economy.

In line with government goals, the Danish Transport Agency has been assigned to administrate a fund for research activities and demonstration projects on energy-efficient transport. The largest single grant of the first round was given to the project 'Test-an-EV', where 300 EVs are tested for daily use by 2400 families during certain time periods [89]. The partner company for the project is Clever and the test is expected to reveal driving and charging patterns as well as user experiences with EVs. Another large-scale project is named EDISON (Electric vehicles in a Distributed and Integrated market using Sustainable Energy and Open Networks). The project uses the island of Bornholm as a full-scale laboratory to investigate market solutions, electricity network configurations and interaction between energy technologies for EVs [90]. The citizens of Bornholm also participate in the smart-grid project 'EcoGrid EU', and results are exchanged between the two projects [91]. Apart from the island of Bornholm, Copenhagen municipality should also be put forward as a major actor since it is, like Bornholm, part of several EU research and demonstration projects. Essential to all those projects is also the cooperation with Danish universities like DTU that are part of multiple projects.

Industry Position

Denmark is one of the countries where new business models with regard to electric mobility are being implemented. Such companies dedicated to deployment, service systems and infrastructure for EVs are by some addressed as electric mobility operators (EMOs). Central EMOs in Denmark are, for example, Better Place Denmark (owned by Better Place Global with Dong Energy as minority stakeholder), ChoosEV (which is now also called Clever – owned by the energy companies SE, SAES-NVE and the car rental company SIXT), CleanCharge and Clear Drive [92–95]. Better Place, in particular, has received worldwide attention for their business model that, among other features, relies on battery switching stations to overcome the range problem connected to EVs. Clever has also received attention owing to the largest BEV trial within the EU in which 1600 Danish families have so far participated within a period of 3 years [96, 97]. Clever is building up a national charging network and among slow charging stations wants to reach 350 fast charging stations by 2015. An important network is the Danish Electric Vehicle Alliance, which is a trade association for the EV industry in Denmark, formed in 2009 by the Danish Energy Association. The Alliance has initiated projects on standardization and roaming within the charging infrastructure and has recently prepared a long-term EV strategy [26]. Members range from electric distribution and utility companies over the automotive industry to research institutes and smaller projects on EV technology.

1.2.4   Norway

Goals

The EV network elbil.no has a target of reaching 100 000 EVs by 2020. An even more ambitious industry vision is raised by Energi Norge to reach 200 000 BEVs and PHEVs by 2020. The government regularly releases its 10-year plan for development in the transport sector. The latest plan, spanning from 2010 to 2019, emphasizes the environmental impact of the transport sector and goals for limiting GHG emissions. The goal is to limit emissions from transport by 2.5–4.0 million tons of CO_2 equivalents in 2020 according to continuation of the current development in the sector [98]. The country has also set the target to achieve an average emission level of 85 g/km CO_2 in terms of total new vehicle sales by 2020 [99].

Policy Instruments

In order to reach its goals, the Norwegian government encourages the purchase of EVs in various ways. Noteworthy here is that BEVs currently are relieved from the registration tax (also sometimes called one-time tax or import tax) as well as valued-added tax (VAT) and have a much lower annual tax (10–20% that of ICE-propelled vehicles) [100]. These measures are guaranteed until 2017 as long as no more than 50 000 such cars are on the roads [101]. The current government has even preliminary plans to continue them at least until 2020 [102]. BEVs are further relieved from parking fees at public parking lots, road pricing or congestion charges, charges on ferries (but the driver has to pay) and are often allowed to drive in bus lanes that are otherwise reserved for public transport [99]. Also, in Oslo and other areas, most public charging spots are free to use for owners of BEVs.

 Another actor to mention here is the public funding programme Transnova that is currently among other initiatives funding fast charging stations across the country. The agency also funds various other projects aiming at reducing GHG emissions from the transport sector (e.g. trial or pilot programmes). The Norwegian Research Council runs a funding programme called RENERGI with the objective of ensuring environmentally friendly and economic development of the energy infrastructure, including transport solutions.

Industry Position

Norway is, or has been, home to several EV-related start-up companies, among them the car manufacturers Think and Reva as well as the car sharing company MoveAbout. Unfortunately, Think has not yet been able to restart production after its latest bankruptcy in 2011. Furthermore, Norway has active industry associations around EVs that strongly support further developments.

1.2.5   Nordic Comparison

Looking at the overall Nordic perspective, it becomes apparent that there are large differences in how the countries try to support the deployment of electric powertrains. Especially striking is the significant policy gap that exists in Sweden, where the government set the goal of achieving a fossil-fuel-free independent transport sector by 2030 as well as an industry vision of 600 000 BEVs and PHEVs by 2020, but few policies suggest such a development. Instead of deployment, Sweden and, to a lesser extent, Finland have focused on R&D, annual vehicle tax definition reform and demonstration projects but have not yet made the link to actual deployment of EVs. Norway and Denmark, however, have had a more entrepreneurial policy approach, through actively supporting new start-ups while at the same time giving generous tax exemptions to customers for market uptake. However, taking into account the slow renewal rate of vehicle fleets, one can argue that in all countries the number of EVs on the street still lag behind the ambitious goals set forward. Table 1.1 summarizes existing policy frameworks across the four countries in terms of economic, regulatory and cognitive/normative governance mechanisms [103].

The range of policy measures results in different price tags across the Nordic countries, which is exemplified here in Figure 1.5 by using the BEV Nissan Leaf and the fuel-efficient diesel-driven Golf BlueMotion 1.6 TDI (based on exchange rates from June 2012). The figure solely focuses on initial prices at the point of purchase and, hence, does not include operational costs or benefits. The price information is gathered from the original equipment manufacturers' (OEMs') web sites and then combined with the policies that exist in the Nordic countries at the point of sale. It can be clearly seen that BEVs will have a hard time competing in Finland and Sweden given current governance regimes. Even though the BEV is likely favourable in terms of operational costs, it will be difficult to close the existing cost gap within a reasonable investment time frame.

1.3 Scenarios and Environmental Impact Assessment

On the basis of existing EV-related policy targets, this section will elaborate two simple future scenarios. The primary variable in the two scenarios is the rate of market uptake of BEVs and PHEVs. This variable will be specified relying on existing market uptake scenarios focusing on Europe that were identified in a literature review. It becomes apparent that there are quite large differences between those reports and studies [1, 104–107].

In terms of annual vehicle sales percentage, BEVs range between 1 and 12% in 2020 and between 11 and 18% for 2030. In the same way, PHEV and REV combined can be found to be between 4 and 8% in 2020 and between 41 and 66% in 2030.

In terms of total car fleet percentage, BEVs range around 0–1% in 2020 and 3–7% for 2030. In the same way, PHEV and REV combined can be found to be between 0 and 1% in 2020 and between 15 and 26% in 2030.

Owing to the different varieties in the scenario studies found, we decided to consider an incremental as well as a breakthrough scenario largely based on an existing study written for the European Commission [1]. At one end, we hence consider an incremental growth outlook of EV developments given a continued business-as-usual governance regime. This *incremental growth scenario* assumes an 18% vehicle fleet share by 2030 for PHEVs, REVs and BEVs combined. The assumptions for this scenario are as follows:

- battery improvements lack substantial breakthrough;
- lack of coordinated and long-term policy support;
- only limited public acceptance for EVs;
- ICE technology will achieve EU transport targets for 2020, which gives OEMs less incentive to push for EVs in the near future [107].

Table 1.1 EV policy frameworks across the Nordic countries.

	Finland	Sweden	Denmark	Norway
EV targets (Gov. or Ind.)	O No specific EV target	X Industry: 600 000	O No specific EV target	X Ind.: 100 000– 200 000 by 2020
Currently registered BEVs and PHEVs	c. 60	c. 1000	c. 1300	c. 9200
Economic				
VAT exemption	O	O	O	X Norway exempts BEVs from VAT
Registration tax	X The registration tax is adjusted according to CO_2 emissions	O A registration tax does not exist in Sweden	X BEVs are exempted from registration taxes	X BEVs are exempted from registration taxes
Annual vehicle tax reform	X	X	X	X
Company car tax reform	n/a	X	n/a	X
Direct subsidy	O	X Subsidy for super green cars	O	O
Research programmes	X	X	X	X
Demonstration programmes	X	X	X	X
Tolls, congestion, charging fee, parking fee exemption, etc.	O	O	O	X
Regulatory				
Free public charging access	X Some organizations allow free charging	X Some organizations allow free charging	X Some organizations allow free charging	X
Allowance to drive in bus lanes	O	O	O	X
Priority parking	X	X	X	X
Cognitive/normative				
Demonstration programmes that spread information	X	X	X	X

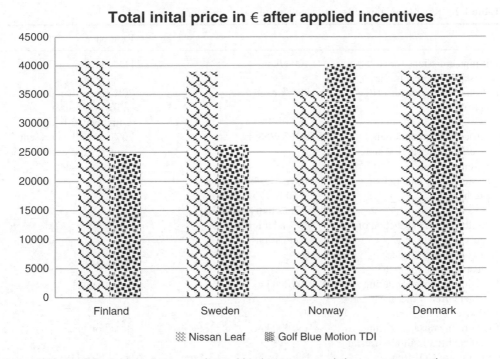

Figure 1.5 Initial price comparison taking into account existing governance regimes.

At the other end, we consider an EV breakthrough scenario, where market share increases rapidly until 2020 and 2030. This *breakthrough scenario* assumes reaching a vehicle fleet share of 33% by 2030 for PHEVs, REVs and BEVs combined. In order for this to be possible, we use a number of important assumptions:

- OEM prices for lithium-ion batteries in the case of BEVs continue to decrease to roughly USD 400/kWh in 2020 and to between USD 150/kWh and USD 200/kWh in 2030 [104, 108, 109];
- strong long-term and coordinated policy support;
- strong public acceptance and behavioural changes in transport [110].

Before now focusing on policies on how to achieve those scenarios, we will first focus on the environmental impact of the market uptake options described. The electrification of vehicles is currently being discussed as a major lever for a more environmentally friendly form of transport. Emissions of nitrogen oxides (NO_x) and particulate matter can be avoided locally and climate impact may be reduced if low-carbon electricity is used.

Here, we will estimate the potential effect of the EV scenarios regarding GHG emissions. A life-cycle perspective is used, which means that emissions associated with vehicle manufacturing and maintenance as well as emissions caused by electricity production are considered, in addition to tail-pipe emissions. First, we calculate life-cycle emissions for three typical vehicles in 2030. These results are then combined with the shares for PHEVs (includes also REVs)

Table 1.2 Key assumptions used to calculate life-cycle emissions.

		Reference
Tail-pipe emissions for an '80 g diesel car' in real traffic	100 g/km	[111, 112]
Tail-pipe emissions for PHEV in petrol mode	95 g/km	20% lower than present Toyota Prius in highway driving
Electricity consumption in electric mode (PHEV and all-electric car)	0.16 kWh/km	20% lower than present energy use according to [112]
Emissions from electricity production	160 g CO_2/kWh (50 and 600 g CO_2/kWh in sensitivity analysis)	[113]
Emissions from production of fuels from oil sand	40% addition to direct emissions	[114]
Share of driving distance in electric mode for PHEV	60%	[115]
Total driving distance during vehicle life for diesel car and PHEV	200 000 km	Based on [1]
Total driving distance during vehicle life for all-electric car	150 000 km	Based on [1]
Emissions of CO_2 from manufacturing and maintenance of cars during their life length	Diesel car 3.3 ton PHEV 4.0 ton All-electric car 4.8 ton	[112, 116–119]

and all EVs in the scenarios, to estimate approximate emission changes for the passenger car fleet in 2030.

The three types of vehicles are: an efficient diesel car emitting 80 g/km CO_2 according to the New European Drive Cycle (NEDC), a PHEV with a 50 km electrical range and an all-electric car with a 150 km range. All cars are assumed to be the size of a Volkswagen Golf. The key assumptions behind the calculations are presented in Table 1.2.

Since we analyse the effect of changes in the vehicle fleet we use marginal emissions for 2030 in the calculations. With such a long-term perspective we need to consider both the build margin and the operating margin. The former is caused by the fact that an increase in electricity demand that may be forecasted well in advance will increase the building of new power plants. The latter is the marginal electricity source used with a fixed set of production plants, given an increased electricity demand. We use one of the scenarios developed by Sköldberg and Unger [113], which incorporates climate policies roughly in line with the 2°C target. The marginal emissions in that scenario amount to 160 g CO_2/kWh as an average for the period 2009–2037. Since the carbon intensity is uncertain, we also use two other levels for a sensitivity analysis: 50 and 600 g CO_2/kWh. In a similar way, marginal reasoning is applied to emissions associated with production of fossil diesel. We apply a 40% addition to the direct emissions, which corresponds to producing diesel from Canadian oil sand [114].

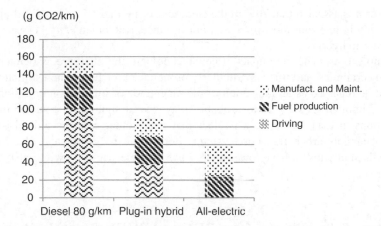

(g CO2/km)

:: Manufact. and Maint.
≈ Fuel production
※ Driving

Diesel 80 g/km   Plug-in hybrid   All-electric

Figure 1.6 Calculated life-cycle emissions 2030 for three types of vehicles (all three sized as a Volkswagen Golf) given a marginal carbon intensity of 160 g CO_2/kWh.

Regarding emissions from manufacturing and maintenance of vehicles, key assumptions used are found in Table 1.2. It is assumed that emissions per car produced are reduced by 40% until 2030, compared with 2005.

Figure 1.6 shows the resulting life-cycle emissions for the three types of cars. At 160 g CO_2/kWh the electric cars are better than the diesel car, although the relative difference is smaller than if only tail-pipe emissions are considered.

We then combine these results with the two scenarios for market penetration: incremental growth with 18% EVs (BEV + PHEV + REV) in the fleet in 2030 and breakthrough with a 33% share in 2030. We assume that the average emissions (according to NEDC) for fossil-fuelled non-plug-in vehicles are 110 g/km in 2030 [120], corresponding to life-cycle emissions of 212 g/km. Furthermore, we assume that biofuels stand for 20% of total energy used for passenger cars, and that they achieve a 70% reduction of GHG emissions compared with fossil fuels.

The resulting changes in life-cycle emissions for passenger cars are shown in Table 1.3. With the middle carbon intensity alternative (160 g), the emission reductions become 6% and 13% respectively. With a very low carbon intensity like 50 g CO_2/kWh the emission reductions amount to 7% and 15%, while a high carbon intensity of 600 g CO_2/kWh GHG gives small emissions reductions. The breakeven level is calculated to be 800 g CO_2/kWh; that is, this is the level which would make emissions unaffected.

Table 1.3 Reduction of GHG emissions in 2030 for different EV market penetration and different CO_2-intensity for electricity production.

	Reduction (%)		
	50 g CO_2/kWh	160 g CO_2/kWh	600 g CO_2/kWh
Incremental growth, 18% of EVs in the fleet	7	6	2
Breakthrough, 33% of EVs in the fleet	15	13	5

In all cases it is assumed that 70% of the electric cars are PHEVs and 30% all-electric cars. This is roughly in line with most forecasts. For instance, Kampman *et al.* [1] assume an 80% share for plug-in hybrids.

All attempts to estimate the impact of new technologies by 2030 are associated with considerable uncertainties, and this is particularly pronounced for EVs. The estimates presented here should be regarded as an indication of the magnitude of impacts on emissions that EVs may have. Although EVs by 2030 probably may give a significant contribution to emission reductions in road transport, it is clear that many other changes will also be needed to reach sustainable urban transport systems. For instance, cycling and electrified public transport will in city traffic have lower energy use than electric cars, while also being more space efficient.

1.4   Future Policy Drivers for a BEV and PHEV Breakthrough

On the basis of existing EV policy targets, this section will elaborate a general breakthrough scenario for strong EV uptake. With this we hope to contribute to an understanding of what an ambitious EV policy goal would actually mean in terms of policy instruments. While doing so, we have gathered existing literature on policy instrument research in the transport sector, or more specifically on hybrids or EVs when available.

First, when analysing technology development and technology shifts it can be valuable to adapt an evolutionary perspective of technical change. From such a point of view, technology develops in technology cycles which can be started by a new 'technological discontinuity' that challenges the old technology [121]. The period in which a new technology challenges the old technology can also be called an 'era of ferment' in which different design options and reactions are triggered around the new technology [121–123]. These options are also referred to as different 'technology trajectories'. Eventually, the era of ferment might end with a new dominant design which becomes the new industry standard since it is the only one that survives the competition for resources [121, 123].

However, new technologies can also fail or have setbacks, and it has to be kept in mind that the development of new technology does not necessarily take place in short time frames but rather necessitates a long-term policy perspective [124]. This can be demonstrated by the fact that EVs were first introduced around the end of the nineteenth century [125]. Also, a new technological discontinuity usually is not alone in challenging an old technology, but itself has many competitors. At the same time, the old technology can react with a strong 'sailing ship effect', in the sense that it improves while it is being challenged [122]. Overall, the technology cycle cannot just be seen from the technology perspective as such, but also has to take into account the overall sociotechnical perspective. The reason for this is that the eventual definition of a new dominant design or technology regime is at least as much shaped by technological, market, legal and social factors as by normative and cognitive frames [121, 123, 126–128].

The evolutionary point of view also stresses that technology usually develops incrementally over time since the development builds on past achievements, ideas and cumulative knowledge [126]. As such, technology is developing along paths which are typically directed at system optimization with reference to the current system logic [126]. Trying to change or influence this direction can be met with a lot of reluctance and prove rather difficult due to sunk investments in existing assets [126, 129]. This, again, can demonstrate how much resistance the Californian Zero Emissions Vehicle policy faced in the early 1990s. Changing the system logic would be a system innovation which would satisfy a societal function in a way that is different from the current sociotechnical system [126]. More precisely, it requires the use of new technology, new markets, new knowledge, new linkages, different rules and roles and major organizational change through, for example, new business models [53, 126].

System innovation can be analytically divided into four different diffusion phases along the S-curve introduced by Rogers, namely pre-development, take-off, acceleration and stabilization [126, 130]. Those different phases have important policy implications when one takes a look at the technology maturity level [131 (p. 397), 132 (p. 407), 133 (p. 9640)].

One important debate in technology and innovation policy is also the question of whether policies should be technology specific or general [134]. Much of that is related to the evolutionary perspective of nurturing both variation and selection [126]. From the selection point of view one can argue that technologies need specific policies which directly interfere with the dynamics of technical change and try to make one path more attractive than others. This is especially necessary if one tries to achieve change on the scale of system innovation in a relatively short time frame. However, these need to be embedded in generic or 'technology neutral' policies, which develop a variety of technology options to be able to select from [126]. Both types of policies have their pros and cons and each will differ according to the technology at hand and the technology's maturity level [126, 135]. What is more important, however, is to give a long-term and clear perspective as a meaningful context for industry and other actors' investment decisions [126].

To incorporate the mentioned multidimensional aspects, a TIS perspective is being adopted which has its strengths in seeing innovation from a systems perspective surrounding the technology. The TIS framework has been adopted by major institutions such as the OECD, the European Commission, UNIDO as well as different Nordic institutions such as the Nordic Council and the Swedish agency Vinnova [132 (p. 407)]. In the literature, a TIS is being defined as 'a network or networks of agents interacting in a specific technology area under a particular institutional infrastructure [e.g. norms and regulation] to generate, diffuse, and utilise technology' [131, 136–138]. The TIS at its heart has a system structure which consists of actors, networks, institutions and artefacts [132 (pp. 408, 413), 138 (p. 817), 139 (pp. 629–630)]. Apart from that, several crucial system processes have been identified and modified in past years [131, 137, 138]. One recent version consists of *entrepreneurial activities, knowledge development and knowledge diffusion, positive external effects, resource mobilization, guidance, market creation, creation of legitimacy and materialization*. Some of these interactive processes need to be addressed by, for example, policy makers at the same time in order to allow reinforcement, feedback mechanisms or complementary action. Also, these processes cannot be seen as disconnected from the system structure and the spatial location of the TIS even if many supply chains are global today. Apart from that, these processes depend heavily on the stage of technology development according to the stages in the 'S-curve'.

Looking at our selection of countries, it is quite possible that Norway and Denmark, for example, are at a different phase of development for their national EV TIS and that in Sweden and Finland, for example, the TIS is still very much facing resistance from the incumbent TIS based around the ICE. After having set up those analytical categories, the following subsection will show policy options that have been identified as potentially supporting an EV TIS. The main focus in the following will be on the mentioned system processes.

1.4.1 Entrepreneurial Activities

Both Norway and Denmark have several companies that have been offering EVs as OEMs or offering EVs in a business model in the form of mobility services. In the case of automobile OEMs, some new EV manufacturers like Think have had mixed results, which is at least partly due to the high entry barriers in the automotive industry [122]. Other start-up companies, like Better Place, Clever and MoveAbout, are slowly starting to become more economically viable. All in all, it is essential to make resources (not just monetary) and knowledge (venturing process, lawyers, marketing, etc.) available for entrepreneurs [140, 141]. This will help to mitigate the real or perceived risks involved of being an entrepreneur and perhaps leaving a secure job [142]. Hence, it is necessary to not design innovation policy instruments only with the known and established actors in mind, but also to account for actors that do not yet exist or for those that are too small to organize their interests [81].

In a breakthrough scenario it is vital to overcome path dependencies often inherent when dealing with established actors and technologies [128, 142]. This makes entrepreneurs that challenge existing technology trajectories a key stepping stone, and there needs to be a good balance between policies supporting entrepreneurs and incumbents (e.g. as in R&D support) [142]. Also, in an early stage of technological development, as is the case with electric cars and lithium-ion batteries, start-ups and entrepreneurs are essential for experimenting around the new technology options and probing ways to commercialize new knowledge [11, 143, 144].

Without commercialization and finding functioning business models, new technologies will not have any value [145–147]. This function should receive special attention in countries with 'big business' bias, like it has been in parts historically found in, for example, Sweden [148–150]. Building up an entrepreneurial environment is essentially also a long-term process that requires patience – in much the same way as it can take several years to find a working business model [11, 150]. Some breakthrough recommendations for this system process hence include:

- Inclusion of entrepreneurial firms in existing government-funded R&D, pilot and demonstration programmes.
- Matching funds and loans for new business ventures.
- Incubator parks, shared office space, shared testing facilities (e.g. like Innovatum or TSS in Sweden) should be more directly supported and increased where reasonable [151].
- Legal and business developing support is perhaps even more important than monetary support for some entrepreneurs as they might lack the necessary business skills and network capital.
- A venture capital fund that is initially matched by government funds could be an interesting instrument if there is a lack of start-up finance in the EV sector [11]. This has been successfully practised in countries like Israel and New Zealand to get investor interest and reduce some of the risk connected with high-tech start-ups.
- Effective evaluation of supported entrepreneurs, in much the same way that is practised by venture capitalists.

1.4.2 Knowledge Development and Knowledge Diffusion

Universities, research networks, pilot projects and demonstration projects are essential to build up the knowledge base in the early stage TIS. On a global level, public-funded research, development and demonstration spending on EVs and PHEVs increased from USD 265 million in 2003 to USD 1.6 billion in 2010 [152]. There have been several European-wide programmes of that kind financed by, for example, the European Investment Bank and the EU's Seventh Research Framework Programme (FP7), as well as several Interreg programmes between countries [1]. Also, in the Nordic national context, there have been several public–private pilot and demonstration projects and some are still ongoing. Owing to the fact that there are still important research efforts to be made when it comes to, for example, battery development or business models, there is a need to keep up such programmes at least in the coming 5–10 years [152].

Also, networks created through research and demonstration programmes can help to build up a national or Nordic knowledge base [61]. This, in turn, helps in creating research, development and demonstration partnerships, industrial partner investments and good practice exchange.

1.4.3 Positive External Effects

Through developing a knowledge base and knowledge networks, supporting entrepreneurs and similar measures, opportunities are created that lead to knowledge spillovers in and between industries [141, 153]. These opportunities can be seized by entrepreneurs that can combine this knowledge in a new way. This, in turn, nurtures positive feedback cycles and helps the industry and the economy to grow. Also, those feedbacks will force incumbents to reconsider their own strategic position in the industry and its value chain [149]. Creating positive externalities that cannot entirely be covered by patents is also an argument for the government giving matching funds and subsidies for start-ups and demonstration projects.

1.4.4 Resource Mobilization

Developing EV drivetrains and infrastructure has usually been helped by governments with R&D support. Having public research programmes that sponsor 25–50% of research efforts made by companies in this area actively encourages OEMs to invest in drivetrain or battery development [2]. Similar efforts have been and can be done to provide matching funds for other pilot and demonstration projects.

It is also interesting that, in the case of Sweden, industrial partnerships have been established to push and commercialize the PHEV technology [53]. In this case, Volvo and Vattenfall together financed the development, making Vattenfall one of the few utilities that directly invested in EV technology [154]. An interesting option is to more strongly support venture capital funds in general or start new funds where public funds would only be used in the beginning to attract further investors to the fund. This could be especially important in light of the ongoing consequences of the financial crisis and due to the heavy reliance of regional small and medium enterprises (SMEs) on traditional bank loans [140]. In Sweden, it has been shown that it is a general problem to generate spinoffs from university research in more regional areas, particularly in the case of knowledge-intensive SMEs [140].

1.4.5   Guidance

Across the globe, several national development plans and road-maps for EVs do exist. If all of those were to be achieved, 1.5 million PHEVs/EVs would be sold by 2015 and 7 million by 2020 [155]. OEMs have so far not had the same level of production capacity that would be necessary to reach those targets [155]. Overall, there is a need for national and supranational road-maps and coordination that specifies goals in the national or, for example, Nordic context. Regional and local authorities need to translate those national goals into concrete local goals.

Apart from national road-maps, an important issue with new technology is standardization. This, on the other hand, limits the extent to which entrepreneurs can experiment with the new technology and it could also represent an entry barrier. However, common plug and charging standards are also a crucial element for a further breakthrough of EVs, as different standards create disincentives [156]. A European-wide standard is expected for 2012, but globally not before 2017 [1]. In this area, perhaps a common Nordic standard would be a good start for further market uptake.

In a similar vein, it is necessary to reform current fuel standards in the EU since the increasing availability of alternative fuels misguides customers. Hence, harmonized accounting and assessment methodologies are needed to understand the well-to-wheel emissions of EVs compared with other technologies [1]. Similarly, common efficiency or energy consumption standards could be used. Harder regulations on average fleet performance will force car manufacturers to get EVs onto the market, perhaps by having conventional vehicles subsidize new ones. Using such standards in common labelling schemes would hence be the next step to not only improving information on CO_2 per kilometre, but also costs per kilometre [18, 157]. What is of utmost importance when dealing with new technologies is also to create a long-term policy environment that reduces risks and manages expectations for companies and investors [18, 81, 142].

1.4.6   Market Creation

As we have seen in Figure 1.5, a mid-sized EV's initial investment is still substantially larger than the average mid-sized ICE. The higher initial investment cost of EV technologies compared with conventional ICEs suggests that, currently, market creation is still a key barrier in the technological innovation system and that a policy framework must include an arsenal of long-term and short-term economic incentives to bring down the initial cost.

Recent studies focusing on total cost of ownership and learning curves have shown that without strong policy support it can take several years and possibly decades until PHEVs and BEVs will break even with HEVs or ICEs [158–163]. Those studies, however, have mostly been conducted in countries like Germany, the Netherlands, USA, and Japan or have taken the EU average, which results in lower initial tax levels when it comes to general car ownership compared with countries with high registration taxes like Norway or Denmark. Also, such studies have some inherent uncertainties when it comes to battery price development, battery densities, the choice of battery technologies and the future electricity and oil prices. Furthermore, in such studies, the operational cost advantages of an EV like, for example, lower fuel costs, lower maintenance costs and lower insurance costs are more difficult to capture;

hence, an important aspect of EV ownership is being missed [2]. In that context one major problem is that customers are reluctant to take into account the total cost of ownership over a longer time frame and typically expect a payback within 3–5 years [2, 18, 155].

In theory, the economic incentives needed can be given before, during or after purchase, they can be designed as a one-time or recurring payment and they can be technology neutral or technology specific [22]. Recently, several economic incentives have been applied throughout Europe, among them tax reduction on sales price, tax reduction after purchase, pure subsidy, scrapping scheme, feebate system, reduction of annual vehicle tax, reduction of registration tax, increased fossil fuel tax, differentiated congestion charges and parking fees, joint or public procurement, subsidies for installing charging infrastructure, quotas for OEMs or CO_2 certificates [1, 2, 22].

Among those incentives, the literature suggests that direct tax reductions are effective, more practical and more appreciated than other instruments by the customer if they are applied at the time of purchase and if directed at the customers instead of subsidizing car dealers [17, 22, 164, 165]. These would, for example, be a reduction of the registration taxes and/or VAT as is, for example, applied in Norway and partly in Denmark. Similarly, direct subsidies instead of tax reductions are also valued by the customer, but the practicality depends a lot on the system that is used. Feebate or bonus malus systems are also accepted by the customers, but here success depends a lot upon how the system is set up [17,157]. For example, if the feebate system is set up stepwise instead of as a gradual linear system, important improvement possibilities will be missed [18]. Likewise, the pivot point of the feebate system should be sufficiently low. The overall problem, however, is that these sorts of incentives also potentially favour high-income groups in society who can or could have bought more expensive environmentally friendly cars anyway [165–167]. However, if, for climate reasons, the priority is to increase the market share of EVs, then the free-rider phenomonen might be a necessary risk.

Tax rebates after the purchase for deduction in income tax or the reduction of the yearly vehicle tax have been found to be less effective or less practical for customers [22]. One of the reasons for this is again that consumers are taking operational costs less into account and that the yearly vehicle tax is relatively low in most countries. However it has been shown that the gas price, which is connected to the level of fuel taxes, had a large impact on, for example, hybrid sales in the USA [165]. Through modelling higher fuel taxes it has also been shown that this increases shares of HEVs and BEVs as well as reduces or at least stabilizes total car fleet size [168]. CO_2-based fuel and yearly vehicle taxes have also some published successes [169].

Having exemptions or reductions for congestion charges, road and ferry tolls, road pricing and parking fees have also proved to be a useful economic instrument in, for example, London, Stockholm and in major municipalities in Norway [1, 170].

In line with tough climate goals, a one-time scrapping scheme could also be considered in order to accelerate the replacement of the current vehicle fleet ('scrappage for replacement') [17, 18]. This could be necessary since in Sweden, for example, almost 50% of the vehicle fleet's emission are caused by cars that are 10 years and older [17, 171]. A recent review of scrapping schemes showed that, overall, old cars were traded in for smaller more fuel-efficient vehicles [167]. However, one has to keep in mind the emissions during other life stages of a car [17]. It is thus necessary to make sure that one of the primary conditions for scrapping schemes is that only high-performing environmental cars are being used as the substitute (e.g. in line with the super green car definition in Sweden).

In general, it is important to realize that transport-related economic instruments interact with each other and can be very dynamic when combined and, in turn, have a significant impact on the willingness to pay of consumers [172]. For example, an increased fuel tax combined with an annual vehicle tax based on CO_2 has a larger effect on willingness to buy than when implemented individually. Also, instruments will differ in terms of their short-term and long-term effectiveness.

The economic instruments applied should be adaptable or revised according to learning curves when it comes to, for example, battery development. Policy makers need to closely monitor costs and technology developments and adapt policy schemes accordingly as economics of scale kick in [1, 2]. It should be argued that the instruments are phased out after EVs have reached a certain market share or when battery prices have reached a certain policy target. Also, subsidies can create rebound effects where total passenger transport increases; as mentioned before, however, this could be regulated through road pricing and similar instruments [1]. It has also been shown that the rebound effect, at least within the transport sector, is not as significant as often suggested and is also limited by, for example, time constraints [18].

In all the four simplified stages of technology development, government policies should consider taking into account the potential markets of such vehicles. This will ask the question of which specific market a policy is created for – in much the same way that companies differentiate their business model according to customers or markets [142, 173]. Examples for EVs here are different markets for private customers, public entities, organizations with fleets and car pools and companies that typically lease cars. This differentiation is especially important in countries like Sweden, for example, where company cars make up more than 50% of new car sales. Hence, another important component in creating a market for EVs is different public or joint procurement initiatives. Here, the procurement program by Stockholm can be mentioned as an example which organized a joint procurement initiative for 6000 EVs [73].

Based on the information gathered, we conclude that in a breakthrough scenario which tries to achieve very ambitious goals the following arsenal of instruments can be applied:

- In line with other environmentally friendly cars, PHEVs and BEVs could benefit from a reduced or exempted VAT. This would put technologies that are still at an early market stage near established technology in terms of initial price. A reduction or exemption of VAT is an effective instrument for the introduction of new automotive technologies.
- Instead of a VAT exemption, a feebate (or bonus malus) system with an ambitious pivot point like, for example, 95 g/km CO_2 that gradually moves towards e.g. 50 g/km CO_2 during a 5–10 year time frame is an effective option [17]. The argument to use this policy is that it is a 'cheaper' option for the government since it is potentially revenue neutral. Furthermore, it is also technology neutral and provides a long-term investment environment.
- A scrapping scheme for cars that are older than 10 years in order to accelerate the replacement of the existing car fleet. The new car should at least manage 50 g/km CO_2 (which is in line with the current super green car rebate in Sweden) or a similar threshold according to a well-to-wheel calculation. In order to avoid free riders, a number of preconditions should be established. The incentive should not be major monetary-wise, but rather a complement to, for example, an existing feebate system (the ad hoc programmes after the financial crisis were around €3000) [167]. The scrapping scheme could also be used to support other low

CO_2 transport modes through vouchers for cycling, train travel or collective transport. This would also help to reduce total vehicle fleet size. Similarly, it has to be evaluated if upgrading old vehicles with new technology would be possible. EVAdapt in Sweden is a company that offers such services.

- Even though it is less accepted politically, increasing the fuel tax and annual vehicle tax (based on CO_2 content of the fuel) is effective. This could also include a minimum price tag so that entrepreneurs can count on a minimum gasoline price for their business models. Such an increase in prices should be phased in gradually.
- The introduction of congestion charges in major cities that also reflect CO_2 emissions in a car's life cycle is an effective mechanism to improve local environmental conditions in cities, but also provides the option of mitigating rebound effects.
- A compulsory labelling scheme that shows cost per kilometre as well as CO_2 per kilometre based on, for example, a well-to-wheel life cycle could be an improvement to existing labelling schemes. A study focusing on hybrid sales in Switzerland has shown that labels affect automotive purchase decisions [157].
- If higher fuel taxes and other mentioned incentives are implemented, lower taxes for low-income groups (e.g. income tax) should be implemented to not disproportionately harm vulnerable groups in society.

1.4.7 Creation of Legitimacy

Arguably, public acceptance and legitimacy are still huge problems when it comes to this technology trajectory, since misunderstanding and misinformation are common both in terms of what EVs can achieve and what they cannot achieve given current technology performance. This requires more information campaigns and possibilities to come into contact with the new technology in, for example, trial programmes.

A general problem in this regard is also the fact that most customers do not consider total cost of ownership when they are purchasing a vehicle [2]. Hence, governments should guide customers by introducing clear labels that take into account the total cost of ownership.

Also, the electrification of transport is highly dependent on decarbonization strategies in the power sector [174, 175]. Only this will give it legitimacy and acceptance in the long term.

Supporting a breakthrough in PHEV and BEV technology can only be one of several measures needed in the transport sector to reach the climate targets. An important factor is also the support of other transport alternatives and modes, as well as behavioural changes [110, 176].

1.4.8 Materialization

Materialization addresses the development of the physical products, factories and infrastructure [153, 177]. Here, crucial elements also are demonstration projects, pilot projects and R&D programmes that provide matching funds for developing the physical infrastructure that is needed. Institutional alignment is also needed to facilitate the charging infrastructure for PHEVs and BEVs.

1.5   Results and Conclusion

Looking at the current policy measures and ambitions in the Nordic countries, it is interesting to acknowledge that it has not been Sweden, the country with the largest automotive industry and the goal of a fossil-fuel-free transport, that has engaged most aggressively with the BEV/PHEV technology. Instead, countries like Norway and Denmark are leading policy developments and have also been home to some of the most innovative business models in the area. This seems to strengthen the idea of path dependencies inherent in the arena of policy, industry and other parts of sociotechnical systems [127–129]. Industry in Sweden, while in-house engaging with electrified powertrains, has been cautious about the right moment to commercialize the technology [178]. This is partly explained by the fact that it requires considerable investment to create new vehicle platforms while at the same time receiving ambivalent policy signals about long-term support mechanisms and having sunk investments in existing vehicle platforms.

What is also apparent is that the most successful country in terms of EV deployment, Norway, is the country that uses the full arsenal of governance mechanisms (economic, regulatory and cognitive/normative) and has guaranteed this policy framework at least until 2017 [103]. The difference becomes more apparent between Denmark and Norway, which both have very strong economic incentives when it comes to initial investment, as can be seen in Figure 1.5. However, Norway facilitates day-to-day EV usage much more through operational economic incentives and regulatory measures that are in place. This saves time and money in operation on top of the favourable initial investment incentives. Denmark might not yet have the high EV sales that were originally anticipated, but it has done important groundwork, especially when it comes to charging infrastructure, cognitive prerequisites (largest EV trial programme in Europe) and initial investment incentives. Finland, at this stage, has not yet prioritized EV deployment even though it does have the necessary industrial base that participates in EV technological innovation systems in other countries, including other Nordic countries.

Looking at the environmental impact of our scenarios, the life-cycle analysis performed indicates that electric cars may by 2030 reduce GHG emissions from passenger cars by up to 15% compared with a reference scenario without any electric cars. The estimates presented in this chapter should be regarded as an indication of the magnitude of impacts on emissions that EVs may have. Although EVs by 2030 may thus give a significant contribution to emission reductions, it is clear that many other changes will also be needed to reach sustainable urban transport systems. For instance, an increased share for cycling and (electrified) public transport will be needed in cities. These modes of transport have even lower energy use than electric cars and are more space efficient.

To reach the existing ambitious climate goals in the transport sector, a number of general breakthrough policy recommendations for BEVs/PHEVs have been given in this chapter. To implement those policies, some Nordic governments have to shift from path-dependent, incremental-change types of policies towards entrepreneurial policies. This includes both support to start-ups and incumbents on the OEM side, but also a clear long-term and short-term policy arsenal to take into account the different development phases of a TIS according to the S-curve. In this regard, it seems prudent to also differentiate between governance mechanisms that interfere at the initial purchase decision and mechanisms that focus on day-to-day operational usage of an EV.

To accelerate developments it seems timely, effective and economic for governments to implement a feebate system. That system could have a pivot point of 95 g/km CO_2 that gradually moves towards 50 g/km CO_2 until 2020 at the latest. On top of that, a scrappage scheme is an interesting option that would accelerate vehicle fleet renewal. This should be done upon ambitious emissions requirements like, for example, the 50 g/km CO_2 which is in line with the Swedish super green car incentive. Instead of trading the old car for a new car, the scheme could also be used to obtain a voucher for collective transport usage, train travel, technology upgrade of the old car or the purchase of bicycles. While the feebate system would be the long term and technology-neutral policy signal, it is very likely that EV power trains will also need a short- to mid-term dedicated policy incentive like, for example, a direct subsidy. To supplement the economic instruments and raise awareness of the total cost of ownership of a car, labelling schemes should be compulsory (with information on CO_2 emissions as well as estimated cost per kilometre).

Acknowledgements

This chapter was made possible through the support of NORSTRAT (Nordic power road map 2050: Strategic choices towards carbon neutrality). This is a 4-year long project which has financing from Nordic Energy Research (NER) (www.nordicenergy.net), where it is part of the larger research programme Sustainable Energy Systems 2050.

References

[1] B. Kampman, H. van Essen, W. Braat, M. Gruenig, R. Kantamaneni and E. Gabel (2011) Impact analysis for market uptake scenarios and policy implications, CE Delft, Delft, April.

[2] B. Kampman, W. Braat, H. van Essen and D. Gopalakrishnan (2011) Economic analysis and business models, CE Delft, Delft, April.

[3] K. Yabe, Y. Shinoda, T. Seki, H. Tanaka and A. Akisawa (2012) Market penetration speed and effects on CO_2 reduction of electric vehicles and plug-in hybrid electric vehicles in Japan. *Energy Policy*, **45**, 529–540.

[4] EEA (2010) Emissions share by sector in EU27, 2010. European Environmental Agency.

[5] SSB (2012), Tabell: 08940: Klimagasser, etter kilde, energiprodukt og komponent, http://statbank.ssb.no/statistikkbanken/selectvarval/Define.asp?MainTable=UtslippKlimaEkvAktN&SubjectCode=01&ProductId=01.04&nvl=True&mt=0&pm=y&PLanguage=0&nyTmpVar=true (accessed 21 November 2012).

[6] Statistics Finland (2011), Greenhouse Gas Emissions in Finland. 1990–2009. Draft. National Inventory Report under the UNFCCC and the Kyoto Protocol. Submission to the European Union, January.

[7] Trafikverket (2011) Ökade utsläpp från vägtrafiken trots rekordartad energieffektivisering av nya bilar, February.

[8] M. Winther (2012) Danish emission inventories for road transport and other mobile sources. Inventories until the year 2010. Aarhus University, Department of Environmental Science, Danish Centre for Environment and Energy, 24 August.

[9] VTT (2012) *Road traffic emissions and energy consumption in Finland*, VTT Technical Research Centre of Finland, May.

[10] Eurostat (2012), Population data, http://appsso.eurostat.ec.europa.eu/nui/show.do (accessed 29 November 2012).

[11] J. Lerner (2010) The future of public efforts to boost entrepreneurship and venture capital. *Small Business Economics*, **35** (3), 255–264.

[12] European Commission (2007) The Lisbon Treaty, 13 December.

[13] ACEA (2008) Key figures, http://www.acea.be/index.php/news/news_detail/economic_turmoil_hits_vehicle_makers_hard/ (accessed 10 July 2012).

[14] European Commission (2011) Transport: White paper 2011, http://ec.europa.eu/transport/strategies/2011_white_paper_en.htm (accessed 10 July 2012).

[15] F. Creutzig, E. McGlynn, J. Minx and O. Edenhofer (2011) Climate policies for road transport revisited (I): evaluation of the current framework. *Energy Policy*, **39** (5), 2396–2406.

[16] B. Lewis (2012) EU car CO_2 proposals well-tuned-car parts chief. *Reuters*, 14 June.

[17] F. Nemry, K. Vanherle, W. Zimmer *et al.* (2009) Feebate and scrappage policy instruments. Environmental and economic impacts for the EU27. European Commission, Joint Research Centre, Instute for Prospective Technological Studies, Luxembourg:, Office for Official Publications of the European Communities, EUR 23896 EN.

[18] P. Kågesson (2010) Med klimatet i tankarna – styrmedel för energieffektiva bilar, Expertgrupp för miljöstudier. Finansdepartementet, Regeringskansliet., Rapport till Expertgruppen för miljöstudier 2011:1, December.

[19] EEA (2012) CO_2 emissions from new passenger cars, http://www.eea.europa.eu/data-and-maps/data/CO2-cars-emission.

[20] EEA (2012) CO_2 emissions from new passenger cars. Monitoring Report.

[21] Eurostat (2012) CO_2 emissions from new passenger cars, http://epp.eurostat.ec.europa.eu/tgm/table.do?tab=table&init=1&plugin=1&language=en&pcode=tsdtr450.

[22] F. Kley, M. Wietschel and D. Dallinger (2010) Evaluation of European electric vehicle support schemes, Fraunhofer Institute for Systems and Innovation Research (ISI), Working Papers S7/2010.

[23] F. Kley, M. Wietschel and D. Dallinger (2012) Evaluation of European electric vehicle support schemes, in *Paving the Road to Sustainable Transport: Governance and Innovation in Low-Carbon Vehicles* (eds M. Nilsson, K. Hillman, A. Rickne and T. Magnusson), Routledge.

[24] Eurostat (2012) Annual detailed enterprise statistics for industry (NACE Rev.2 B-E), 25 June.

[25] Grønn Bil (2012), Ladbare biler i Norge, http://gronnbil.no/elbiluniverset/kart.php#zoom=4&tr=72.14173187862764,56.444476074218755&bl=55.70293210778397,-30.567242675781245&m=1&regamp;=0 (accessed 15 February 2012).

[26] Dansk Elbil Alliance (2012) E-mobilitet køreplan 2020. En segmenteret markedstilgang er nøglen til at få igangsat en effektiv udbredelse af elbiler, November.

[27] easycharge (2012) Statistik – ELIS, 1 November, http://www.easycharge.se/tj%C3%A4nster/statistik-elis-11315525 (accessed 24 November 2012).

[28] B. Godske (2012) Hvor mange elbiler er der i Danmark i 2020?, *Motorbloggen | Ingeniøren*, 12 March.

[29] hbl.fi (2012) Endast 20 elbilar sålda i Finland i år, *Hbl.fi*, 20 June.

[30] Helsingborg stad (2012) Årets elbilskommun utmanar regeringen, *Mynewsdesk*, 5 July.

[31] S. Nordgren (2012) Elbil inte alternativ nummer ett, *svenska.yle.fi*, 5 March.

[32] Finnish Transport Agency (2011) Transport conditions 2035, February.

[33] Finish Government (2009) Government Foresight Report on Long-term Climate and Energy Policy: Towards a Low-Carbon Finland, November.

[34] Finansministeriet (2011) Skatteuppgörelser i budgetpropositionen för 2012, October.

[35] Finansministeriet (2011) Ändringar inom finansministeriets verksamhetsområde som träder i kraft vid årsskiftet, December.

[36] M. Kosk (2010) Utsläppsfria bilar bör gynnas, *Hbl.fi | Finlands ledande nyhetssajt på svenska*, 2 March.

[37] Finansministeriet (2012) Ändringen av bilskattelagen träder i kraft den 1 april 2012, March.

[38] C.-G. Lindén (2011) Elbilarna strömmar till, men långsamt, *Hbl.fi | Finlands ledande nyhetssajt på svenska*, 31 December.

[39] FMEE (2009) Electric vehicle report, Finish Ministry of Employment and the Economy.

[40] Tekes (2011) EVE – Brochure, June.

[41] Tekes (2012) EVE – Electric Vehicle Systems 2011–2015, http://www.tekes.fi/programmes/EVE (accessed 16 April 2012).

[42] electrictraffic.fi (2012) Electric Traffic Helsinki Test Bed, http://sahkoinenliikenne.fi/ (accessed 18 April 2012).

[43] H.H. Kvisle (2012) Finland i elbilfarta, 16 February, http://www.elbil.no/politikk/603-finland-i-elbilfarta (Accessed 19 April 2012).

[44] yle Nyheter (2012) Elbil inte alternativ nummer ett, *yle Nyheter*, 5 March.

[45] www.evelina.fi (2012) EVELINA – National Test Environment for Electric Vehicles, http://www.evelina.fi/ (accessed 18-April 2012).

[46] eco-urbanliving.com (2012) Eco Urban Living Initative, http://www.eco-urbanliving.com/index.php/about-us.html (accessed 18 April 2012).

[47] Tekes (2011) Sustainable community 2007–2012, http://www.tekes.fi/programmes/Yhdyskunta (accessed 18 April 2012).

[48] Aalto University (2012) Elbilstrenden i Finland börjar med batterier, 2 April, http://chem.aalto.fi/sv/current/news/view/2012-04-02/ (accessed 17 April 2012).

[49] E. Mellgren (2010) Finland tar täten i elbilsracet, NyTeknik, 21 April.

[50] L.A. Karlberg (2010) Valmet, Nokia och Fortum lanserar finska elbilen EVA, NyTeknik, 3 March.

[51] D. Kronqvist (2011) Det ljusnar för finländsk elbil, Hbl.fi | Finlands ledande nyhetssajt på svenska, 5 May.

[52] J. Hållén (2010) Finland inviger toppmodern batterifabrik, NyTeknik, 11 June.

[53] M. Albrecht (2011) Electromobility in Sweden – towards a new dominant business model design? A perspective looking through the eyes of utilities active on the Swedish market, Lund University.

[54] Infrastrukturnyheter (2011) Fortum i storsatsning på elbilar | Infrastrukturnyheter.se, 2 September.

[55] Elforsk, TSS and Power Circle (2010) Förslag till Nationellt Program för Utvärdering av Elfordon och Laddningsinfrastruktur, July.

[56] A. K. Hatt (2012) Fossilfria transporter – vägen till ett långsiktigt hållbart samhälle. Swedish Government, 21 March.

[57] Power Circle, ElForsk and Test Site Sweden (2009) Gör Sverige Till En Ledande Elbilsnation.

[58] H. Sköldberg, E. Löfblad, D. Holmström et al. (2010) Ett fossilbränsleoberoende transportsystem år 2030, Elforsk & Svensk Energi, May.

[59] Svensk Energi (2011) En fossiloberoende transportsektor år 2030 – hur går vägen dit?, September.

[60] Energimyndigheten (2009) Knowledge base for the market in electric vehicles and plug-in hybrids.

[61] A. Lewald (2011) Interview with Swedish Energy Agency.

[62] Miljofordon (2012) Vad är miljöfordon?, http://www.miljofordon.se/fordon/vad-ar-miljobil (accessed 29 January 2012).

[63] Skatteverket (2012) Information om bilförmånsberäkning 2012 och 2013, http://www.skatteverket.se/privat/etjanster/bilformansberakning/2012/info2012 4.71004c4c133e23bf6db80008894.html (accessed 19 April 2012).

[64] Swedish Government (2011) Regeringen inför supermiljöbilspremien, http://www.regeringen.se/sb/d/8756/a/174478 (accessed 29 January 2012).

[65] Swedish Government (2011) Regeringen satsar 200 miljoner kronor på supermiljöbilspremien, http://www.regeringen.se/sb/d/15436/a/183400 (accessed 29 January 2012).

[66] SvD (2012) Populär supermiljöbil tömmer statens premiekassa, SvD, 22 October.

[67] P. Lundgren (2011) Interview with Öresundskraft.

[68] Malmö City (2009) Malmö Plug-in City, http://www.malmo.se/Medborgare/Miljo--hallbarhet/Miljoarbetet-i-Malmo-stad/Projekt--natverk/Projekt/Elfordon/Plug-in-City.html.

[69] Malmö City (2009) E-mobility Project, http://www.malmo.se/Medborgare/Miljo--hallbarhet/Miljoarbetet-i-Malmo-stad/Projekt--natverk/Projekt/Elfordon/E-mobility.html.

[70] Stockholm City (2009) MobilEl. En demonstration av laddhybrider i Stockholm.

[71] E. Sunnerstedt (2011) Interview with Eva Sunnerstedt, Stockholm City.

[72] U. Östermark (2011) Interview with Göteborg Energi, Programme Manager of the Smart Grid Programme.

[73] Elbilsupphandling.se (2011) Upphandling av 6000 elbilar, http://www.elbilsupphandling.se/upphandling-av-6000-elbilar/.

[74] Swedish Transport Agency (2011) Nytt vägmärke visar laddplats för elfordon, http://www.transportstyrelsen.se/sv/Nyhetsarkiv/Nytt-vagmarke-visar-laddplats-for-elbilar/.

[75] M. Alpman (2010) Lättare bygga laddstolpar för elbilar, NyTeknik, 29 November.

[76] Energimarknadsinspektionen (2010) EI R2010:20. Uppladdning för framtidens fordon. Undantag från koncession för laddinfrastruktur.

[77] Energimarknadsinspektionen (2010) Uppdrag till EI om laddinfrastruktur för elbilar, http://www.ei.se/For-press/Aktuellt-fran-inspektionen/Uppdrag-till-EI-om-laddinfrastruktur-for-elbilar/.

[78] Energimarknadsinspektionen (2011) Regeringen föreslår nya regler på elmarknaden, http://www.ei.se/For-press/Aktuellt-fran-inspektionen/Regeringen-foreslar-nya-regler-pa-elmarknaden/.

[79] A.-K. Hatt (2012) Ett hinder mindre på vägen, 4 April.

[80] S. Wade (2012) My interview with NEVS about their plans for Saab, Swadeology, 15 November.

[81] M. Albrecht (2011) Policy and innovation in the clean tech industry. A focus on electromobility, Course paper, March.

[82] TRM (2009) Aftaler om en grøn transportpolitik, Danish Ministry of Transport (ISBN: 978-87-91013-16-4).

[83] Ritzau (2012) Elselskaber frygter politiske besparelser, *information.dk*, 3 February.

[84] Ritzau (2012) Læs her om energiaftalen i hovedtræk, *information.dk*, 22 March.

[85] Dansk Elbil Alliance (2012) Forside | Dansk Elbil Alliance, http://www.danskelbilalliance.dk/ (accessed 15 February 2012).

[86] ENS (2012) Elbiler, http://www.ens.dk/da-DK/KlimaOgCO2/Transport/elbiler/Sider/Forside.aspx (accessed 15 February 2012).

[87] TRM (2011) Et grønnere transportsystem, Danish Ministry of Transport (ISBN: 978-87-91013-95-9).

[88] DMT (2012) SKAT: Registreringsafgift for nye køretøjer – satser, Danish Ministry of Taxation.

[89] testenelbil.dk (2012) Bag om projektet | testenelbil.dk, http://testenelbil.clever.dk/bag-om-projektet/ (accessed:11 July 2012).

[90] Edison (2012) About the Edison project, http://www.edison-net.dk/About_Edison.aspx (accessed 15 February 2012).

[91] EcoGrid (2012) EcoGrid EU, http://energinet.dk/en/forskning/EcoGrid-EU/sider/EU-EcoGrid-net.aspx (accessed 15 February 2012).

[92] L. Borking (2012) Nu kommer gennembruddet for elbilerne – måske, *information.dk*, 1 February.

[93] CleanCharge (2012) Forside, http://www.cleancharge.dk/Forside.html (accessed 11 July 2012).

[94] Clever (2012) CLEVER – vi oplader elbilerne!, http://www.clever.dk/om-clever/# (accessed 11 July 2012).

[95] S. Møller (2012) Firma klar med 150 dele-elbiler i København: vi henter dem, når de løber tør, *Ingeniøren*, 21 November.

[96] ChoosEV (2012) ChoosEV's elbiler runder to millioner kørte kilometer, 17 April, http://www.choosev.com/nyheder/choosev%27s-elbiler-runder-to-millioner-koerte-kilometer/ (accessed 20 April 2012).

[97] CSR (2012) Elbilerne runder 3 mio i Danmark, *Erhvervsmagasinet CSR*, 21 November.

[98] NMTC (2009) National Transport Plan 2010–2019, Norwegian Ministry of Transport and Communications.

[99] Norwegian Government (2012) Melding til Stortinget 21. Norsk Klimapolitikk, April.

[100] H. Seljeseth (2011) Opportunities and challenges with large scale integration of electric vehicles and plug-in-vehicles in the power system, in *Smart Grids and E-Mobility*, Ostbayerisches Technologie-Transfer-Institut e.V. (OTTI), Regensburg, pp. 180–192.

[101] Grønn Bil (2012) Klimaforliket: Elbil-fordelene sikret til 2017, 8 June, http://gronnbil.no/nyheter/klimaforliket-elbil-fordelene-sikret-til-2017-article262-239.html (accessed 9 July 2012).

[102] M. Johansen (2012) Fritt frem for elbilene i kollektivfeltet, *Aftenposten*, 25 April.

[103] M. Nilsson, K. Hillman and T. Magnusson (2012) How do we govern sustainable innovations? Mapping patterns of governance for biofuels and hybrid-electric vehicle technologies. *Environmental Innovation and Societal Transitions*, **3**, 50–66.

[104] Bloomberg (2012) Electric vehicle battery prices down 14% year on year | Bloomberg New Energy Finance, 16 April, http://www.bnef.com/PressReleases/view/210 (accessed 17 April 2012).

[105] M. Book, X. Mosquet, G. Sticher *et al.* (2009) The comeback of the electric car? How real, how soon, and what must happen next, Boston Consulting Group, January.

[106] A. Dinger, R. Martin, X. Mosquet *et al.*, (2010) Batteries for electric cars: challenges, opportunities, and the outlook to 2020, Boston Consulting Group, January.

[107] X. Mosquet, M. Devineni, T. Mezger *et al.* (2011) Powering autos to 2020: the era of the electric car? Boston Consulting Group, July.

[108] G. Duleep, H. van Essen, B. Kampman and M. Gruenig (2011) Assessment of electric vehicle and battery technlogy, CE Delft, Delft, 11.4058.04, April.

[109] H. van Essen and B. Kampman (2011) Impacts of electric vehicles – summary report, CE Delft, Delft, 11.4058.26, April.

[110] J. Anable, C. Brand, M. Tran and N. Eyre (2012) Modelling transport energy demand: a socio-technical approach. *Energy Policy*, **41**, 125–138.

[111] K. Burgdorf (2011) Challenges and opportunities for the transition to highly energy-efficient passenger cars, Volvo Car Corporation, Warrendale, PA, SAE Technical Paper 2011-37-0013, June.

[112] J. Patterson, M. Alexander and A. Gurr (2011) Preparing for a life cycle CO_2 measure, low carbon vehicle partnership. Ricardo, RD.11/124801.5, August.

[113] H. Sköldberg and T. Unger (2008) Effekter av förändrad elanvändning/elproduktion – Modellberäkningar, Elforsk, Elforsk rapport 08:30, April.

[114] A.D. Charpentier, J.A. Bergerson and H.L. MacLean (2009) Understanding the Canadian oil sands industry's greenhouse gas emissions. *Environmental Research Letters*, **4** (1), 014005.

[115] J. Åkerman, K. Isaksson, J. Johansson and L. Hedberg (2007) Tvågradersmålet i sikte? Scenarier för det svenska energi- och transportsystemet 2050, Naturvårdsverket, Rapport 5754, October.

[116] H.-J. Althaus and M. Gauch (2010) Vergleichende Ökobilanz individueller Mobilität: Elektromobilität versus konventionelle Mobilität mit Bio- und fossilen Treibstoffen, Life Cycle Assessment and Modelling Group, Technologie und Gesellschaft, Empa, October.

[117] M.M. Hussain, I. Dincer and X. Li (2007) A preliminary life cycle assessment of PEM fuel cell powered automobiles. *Applied Thermal Engineering*, **27** (13), 2294–2299.

[118] C. Samaras and K. Meisterling (2008) Life cycle assessment of greenhouse gas emissions from plug-in hybrid vehicles: implications for policy. *Environmental Science and Technology*, **42** (9), 3170–3176.

[119] N. Zamel and X. Li (2006) Life cycle analysis of vehicles powered by a fuel cell and by internal combustion engine for Canada. *Journal of Power Sources*, **155** (2), 297–310.

[120] WSP (2008) Bilparksprognos i åtgärdsplaneringen. EET-scenario och referensscenario, WSP Analys & Strategi, Rapport 200825, December.

[121] P. Anderson and M.L. Tushman (1990) Technological discontinuities and dominant designs: a cyclical model of technological change. *Administrative Science Quarterly*, **35** (4), 604–633.

[122] H. Pohl and M. Yarime (2012) Integrating innovation system and management concepts: the development of electric and hybrid electric vehicles in Japan. *Technological Forecasting and Social Change*, **79** (8), 1431–1446.

[123] M.L. Tushman and P. Anderson (1986) Technological discontinuities and organizational environments. *Administrative Science Quarterly*, **31** (3), 439–465.

[124] C. Wilson (2012) Up-scaling, formative phases, and learning in the historical diffusion of energy technologies. *Energy Policy*, **50**, 81–94.

[125] K.G. Hoyer (2008) The history of alternative fuels in transportation: the case of electric and hybrid cars. *Utilities Policy*, **16** (2), 63–71.

[126] M. Arentsen, R. Kemp and E. Luiten (2002) Technological change and innovation for climate protection: the governance challenge, in *Global Warming and Social Innovation: The Challenge of a Climate-Neutral Society* (ed. M.T.J Kok) Earthscan, chapter 4.

[127] T.P. Hughes (1993) *Networks of Power: Electrification in Western Society, 1880–1930*, JHU Press.

[128] G. Unruh and P. del Río (2012) Unlocking the unsustainable techno-institutional complex, in *Creating a Sustainable Economy: An Institutional and Evolutionary Approach to Environmental Policy* (ed. G. Marletto), Routledge, pp. 231–255.

[129] G. C. Unruh (2000) Understanding carbon lock-in. *Energy Policy*, **28** (12), 817–830.

[130] E.M. Rogers (2003) *Diffusion of Innovations*. Free Press, New York, NY.

[131] K. van Alphen, M. P. Hekkert and W. C. Turkenburg (2010) Accelerating the deployment of carbon capture and storage technologies by strengthening the innovation system. *International Journal of Greenhouse Gas Control*, **4** (2), 396–409.

[132] A. Bergek, S. Jacobsson, B. Carlsson, S. Lindmark and A. Rickne (2008) Analyzing the functional dynamics of technological innovation systems: a scheme of analysis. *Research Policy*, **37** (3), 407–429, doi:10.1016/j.respol.2007.12.003.

[133] R.A.A. Suurs, M.P. Hekkert and R.E.H.M. Smits (2009) Understanding the build-up of a technological innovation system around hydrogen and fuel cell technologies. *International Journal of Hydrogen Energy*, **34** (24), 9639–9654, doi:10.1016/j.ijhydene.2009.09.092.

[134] B.A. Sandén and C. Azar (2005) Near-term technology policies for long-term climate targets –economy wide versus technology specific approaches. *Energy Policy*, **33** (12), 1557–1576.

[135] A. Bergek and S. Jacobsson (2010) Are tradable green certificates a cost-efficient policy driving technical change or a rent-generating machine? Lessons from Sweden 2003–2008. *Energy Policy*, **38** (3), 1255–1271.

[136] B. Carlsson and R. Stankiewicz (1991) On the nature, function and composition of technological systems. *Journal of Evolutionary Economics*, **1**, 93–118.

[137] M.P. Hekkert and S.O. Negro (2009) Functions of innovation systems as a framework to understand sustainable technological change: empirical evidence for earlier claims. *Technological Forecasting and Social Change*, **76** (4), 584–594.

[138] S. Jacobsson and A. Bergek (2004) Transforming the energy sector: the evolution of technological systems in renewable energy technology. *Industrial and Corporate Change*, **13** (5), 815–849.

[139] S. Jacobsson and A. Johnson (2000) The diffusion of renewable energy technology: an analytical framework and key issues for research. *Energy Policy*, **28** (9), 625–640, doi:10.1016/S0301-4215(00)00041-0.

[140] B. Berggren and L. Silver (2010) Financing entrepreneurship in different regions: the failure to decentralise financing to regional centres in Sweden. *Journal of Small Business and Enterprise Development*, **17** (2), 230–246.

[141] J. Lerner and J. Tåg (2012) Institutions and venture capital, *SSRN eLibrary*, January.

[142] R. Wüstenhagen and E. Menichetti (2012) Strategic choices for renewable energy investment: Conceptual framework and opportunities for further research. *Energy Policy*, **40**, 1–10.

[143] D. B. Audretsch, S. Heblich, O. Falck and A. Lederer (eds) (2011) *Handbook of Research on Innovation and Entrepreneurship*. Edward Elgar Publishing Ltd.

[144] A. Bergek (2012) The role of entrepreneurship and markets for sustainable innovation, in *Creating a Sustainable Economy: An Institutional and Evolutionary Approach to Environmental Policy* (ed. G. Marletto), Routledge, pp. 205–230.

[145] D.J. Teece (1986) Profiting from technological innovation: implications for integration, collaboration, licensing and public policy. *Research Policy*, **15** (6), 285–305.

[146] D.J. Teece (2006) Reflections on 'Profiting from Innovation'. *Research Policy*, **35** (8), 1131–1146.

[147] D.J. Teece (2010) Business models, business strategy and innovation, *Long Range Planning*, **43** (2–3), 172–194.

[148] G. Eliasson (2009) Policies for a new entrepreneurial economy, in *Schumpeterian Perspectives on Innovation, Competition and Growth* (eds U. Cantner, J.-L. Gaffard, and L. Nesta), Springer Berlin, pp. 337–368.

[149] K. Hockerts and R. Wüstenhagen (2010) Greening Goliaths versus emerging Davids – theorizing about the role of incumbents and new entrants in sustainable entrepreneurship. *Journal of Business Venturing*, **25** (5), 481–492.

[150] U. Jakobsson (2011) Interview with the Managing Director of Move About AB.

[151] A. Bergek and C. Norrman (2008) Incubator best practice: a framework, *Technovation*, **28** (1–2), 20–28.

[152] L. Fulton (2011) Electric vehicles: are they a passing fad, or here to stay? International Energy Agency, 2 May.

[153] A. Bergek, S. Jacobsson and B.A. Sandén (2008) 'Legitimation' and 'development of positive externalities': two key processes in the formation phase of technological innovation systems. *Technology Analysis and Strategic Management*, **20** (5), 575–592.

[154] U. Frieser (2011) Interview, Vattenfall, Development Program Manager E-Mobility.

[155] IEA (2011) Technology roadmap: electric and plug-in hybrid electric vehicles, International Energy Agency, June.

[156] S. Brown, D. Pyke and P. Steenhof (2010) Electric vehicles: the role and importance of standards in an emerging market. *Energy Policy*, **38** (7), 3797–3806.

[157] R. Wüstenhagen and K. Sammer (2007) Wirksamkeit umweltpolitischer Anreize zum Kauf energieeffizienter Fahrzeuge: eine empirische Analyse Schweizer Automobilkunden. *Journal of Environmental Research*, **18** (1), 61–78.

[158] C.-S. Ernst, A. Hackbarth, R. Madlener *et al.* (2011) Battery sizing for serial plug-in hybrid electric vehicles: a model-based economic analysis for Germany. *Energy Policy*, **39** (10), 5871–5882.

[159] V.J. Karplus, S. Paltsev and J.M. Reilly (2010) Prospects for plug-in hybrid electric vehicles in the United States and Japan: a general equilibrium analysis. *Transportation Research Part A: Policy and Practice*, **44** (8), 620–641.

[160] G. Pasaoglu, M. Honselaar and C. Thiel (2012) Potential vehicle fleet CO_2 reductions and cost implications for various vehicle technology deployment scenarios in Europe. *Energy Policy*, **40**, 404–421.

[161] C. Thiel, A. Perujo and A. Mercier (2010) Cost and CO_2 aspects of future vehicle options in Europe under new energy policy scenarios. *Energy Policy*, **38** (11), 7142–7151.

[162] O. van Vliet, A.S. Brouwer, T. Kuramochi *et al.* (2011) Energy use, cost and CO_2 emissions of electric cars. *Journal of Power Sources*, **196** (4), 2298–2310.

[163] M. Weiss, M.K. Patel, M. Junginger *et al.* (2012) On the electrification of road transport – learning rates and price forecasts for hybrid-electric and battery-electric vehicles. *Energy Policy*, **48**, 374–393.

[164] P. de Haan, A. Peters and R.W. Scholz (2007) Reducing energy consumption in road transport through hybrid vehicles: investigation of rebound effects, and possible effects of tax rebates. *Journal of Cleaner Production*, **15** (11–12), 1076–1084.

[165] D. Diamond (2009) The impact of government incentives for hybrid-electric vehicles: evidence from US states. *Energy Policy*, **37** (3), 972–983.

[166] A. Chandra, S. Gulati and M. Kandlikar (2010) Green drivers or free riders? An analysis of tax rebates for hybrid vehicles. *Journal of Environmental Economics and Management*, **60** (2), 78–93.

[167] A. Schweinfurth (2009) Car-scrapping schemes: an effective economic rescue policy? IISD, http://www.iisd.org/publications/pub.aspx?id=1260.

[168] M. Kloess and A. Müller (2011) Simulating the impact of policy, energy prices and technological progress on the passenger car fleet in Austria – a model based analysis 2010–2050. *Energy Policy*, **39** (9), 5045–5062.

[169] F. Rogan, E. Dennehy, H. Daly *et al.* (2011) Impacts of an emission based private car taxation policy – first year ex-post analysis. *Transportation Research Part A: Policy and Practice*, **45** (7), 583–597.

[170] M. Börjesson, J. Eliasson, M.B. Hugosson and K. Brundell-Freij (2012) The Stockholm congestion charges – 5 years on. Effects, acceptability and lessons learnt. *Transport Policy*, **20**, 1–12.

[171] SCB (2012) Fordonsbestånd 2011, korrigerad 2012-04-27, April.

[172] S. Mandell (2009) Policies towards a more efficient car fleet. *Energy Policy*, **37** (12), 5184–5191.

[173] U. Dewald and B. Truffer (2011) Market formation in technological innovation systems – diffusion of photo-voltaic applications in Germany. *Industry and Innovation*, **18** (3), 285–300.

[174] R.T. Doucette and M.D. McCulloch (2011) Modeling the CO_2 emissions from battery electric vehicles given the power generation mixes of different countries. *Energy Policy*, **39** (2), 803–811.

[175] R.T. Doucette and M.D. McCulloch (2011) Modeling the prospects of plug-in hybrid electric vehicles to reduce CO_2 emissions. *Applied Energy*, **88** (7), 2315–2323.

[176] F. Cuenot, L. Fulton and J. Staub (2012) The prospect for modal shifts in passenger transport worldwide and impacts on energy use and CO_2. *Energy Policy*, **41**, 98–106.

[177] H. Hellsmark and S. Jacobsson (2008) Opportunities for and limits to academics as system builders – the case of realizing the potential of gasified biomass in Austria, presented at the DIME International Conference Innovation, Sustainability and Policy.

[178] J. Konnberg (2011) Interview with Johan Konnberg (Senior Advisor at the Volvo Car Corporation – Special Vehicles MSS; Commercial and Business Manager for the C30).

2

EVs and the Current Nordic Electricity Market[*]

Christian Bang,[1] Camilla Hay,[1] Mikael Togeby[1] and Charlotte Søndergren[2]
[1]Ea Energy Analyses A/S, Copenhagen, Denmark
[2]Danish Energy Association, Lyngby, Denmark

2.1   Introduction

This chapter describes the current Nordic electricity market and addresses the challenges related to a large-scale introduction of electric vehicles (EVs) with Denmark as a reference point.

An EV will increase electricity consumption for the typical Danish house by 50–60%. If the charging of many EVs primarily takes place during the peak consumption hours between 17:00 and 19:00, this could possibly cause pressure on the energy system and congestions in the grid. By introducing demand response in connection with EVs, congestion challenges can be met even with a very high penetration of EVs. Demand response can, for example, be based on the electricity market's spot prices.

For the end-user to benefit from demand response, an interval meter is needed. Grid companies comprising 50% of all Danish end-users have installed, or will install, new meters within a few years. These meters will be able to record the consumption per hour and thereby make it possible to use price contracts with prices varying per hour (spot prices or critical peak prices), by weekdays/weekends or day/night (time of use).

If EVs are introduced in the spot market, the market set-up is simple and possible today with an interval meter. The retailer can broadcast the electricity prices once a day and the end-user can make a charging strategy for the hours with known prices (12–36 h ahead). The charging strategy can be with a simple clock charging, the cheapest hours can be selected with a local

---

[*]This chapter presents the findings from report 2.3 of the Edison project, 'Introducing electric vehicles into the current electricity markets', and the FlexPower project (www.flexpower.dk).

Grid Integration of Electric Vehicles in Open Electricity Markets, First Edition. Edited by Qiuwei Wu.
© 2013 John Wiley & Sons, Ltd. Published 2013 by John Wiley & Sons, Ltd.

computer system (home automation system) or a form of remote control via a fleet operator could be utilized.

If EVs are to participate in both the spot market and the regulating market, a few more challenges have to be met. Requirements from the transmission system operator (TSO) regarding real-time measurements of the individual unit and minimum bid size make it difficult for EVs to participate in the regulating power market today. Furthermore, there are some challenges with imbalances related to EVs in the regulating market, as the activation of regulating power in one hour can change the predicted charging in a later hour.

Some of these challenges can be met by introducing a fleet operator to aggregate the consumption of a number of EVs and handle their interaction with the electricity market.

It should be noted that the design of the market is not fixed. For example, the Danish TSO has indicated a desire to undertake changes to the regulating power market in order to activate smaller consumption in the market. However, changes might take some time to implement as the regulating power market today is strictly Nordic, while in the future it is planned to be European.

In addition to the wholesale market solutions, there are some challenges in relation to the local grid. Congestions in the local grid have to be taken into account with respect to the behaviour of the end-user. This subject is briefly described within this chapter, but not solved. It is discussed in greater detail in Chapter 3.

2.2   Electricity Consumption by EVs

2.2.1   Typical Consumption of an EV

The electricity consumption for EVs varies depending on the size, make, model and technology involved. Based on a small–medium-sized vehicle, it has been estimated that the average new EV produced within the next 5 years will consume 130–140 Wh/km [1]. This figure is not based on a single particular vehicle, but instead on a theoretical small–medium-sized EV with current motor, battery and charger efficiencies.

The average number of kilometres driven yearly via a traditional personal vehicle is 16 600 km [2]. Given a value of 135 Wh/km, this represents an annual electricity consumption of 2240 kWh per vehicle (owing to a more limited range, today's EVs may drive less kilometres than a typical personal vehicle and would, therefore, have an annual electricity consumption lower than this value). To put this into perspective, the typical Danish family living in a house without electric heating currently uses approximately 4000 kWh per year [3]. As such, for the average Danish house with an EV this represents a 56% increase in annual electricity consumption. For houses with electric heating the percentage increase would be lower, while for apartments the percentage increase would be higher. Assuming 25% of all personal vehicles became electric, this would correspond to over half a million EVs in Denmark, and an increase of total electricity consumption of less than 3%.

2.2.2   Potential Challenges for Electrical Grids

What makes EVs different from the majority of other household electrical usage (lights, washing machine, computer, etc.) is that the actual use of the electricity (while driving) does not take place at the same time as the consumption (while charging). Subject to driving pattern and the type of electrical charger installation, charging for most EVs is anticipated to take 2–8 h for a full battery. Charging can be realized with standard single- or three-phase connections [4].

Depending on when charging takes place, this either represents a significant challenge (need for capacity expansion) or an immense opportunity for electrical power systems (introduction of energy storage in the region). If a significant amount of households all begin to charge their vehicles immediately after they arrive home from work, this will increase electricity demand during what is typically a peak time. This could be problematic for the electrical power system as a whole. If however, the vast majority of EV users could postpone the charging of their vehicles till late in the evening/very early in the morning, then their vehicles would still be fully charged for the following day and charging would occur at an off-peak time.

Main Grid

Based on overall Danish electricity demand, Figure 2.1 illustrates the situation described above. Both graphs in the figure present the hourly national total electricity demand for the day in 2007 that exhibited the highest peak demand hour, namely 24 January. In addition, the graphs also indicate what the hourly demand pattern would look like under two alternative scenarios if 25% of all Danish passenger vehicles were electric. These scenarios are:

● immediate charging (the majority of EV owners will simply plug in their vehicles when they arrive home from work);
● market-based/fleet operator managed charging (price signals or a fleet operator largely dictate the optimal time to charge the EV).

The lighter portions in the figure are the hourly demand for 24 January 2007 (the day in 2007 that had the highest peak hourly demand). Meanwhile, the darker portions represent additional demand corresponding to the electrification of 25% of current passenger vehicles.

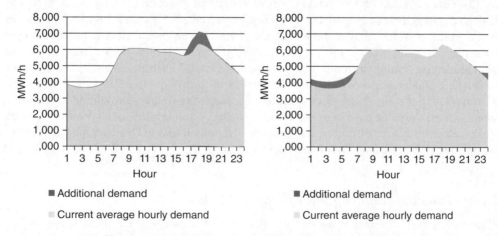

Figure 2.1 The 2007 peak electricity demands for Denmark.

The first graph illustrates the situation without any demand response and the second with the assumed intelligent automated demand response. The left portion highlights the fact that if the majority of charging is done shortly after arriving home from work the introduction of EVs will lead to an increase in peak demand of 12.5% relative to the 2007 peak workday hour. Such an increase in peak demand would imply a more congested transmission grid and higher prices; however it does not necessarily threaten the security of supply.

In scenarios with effective demand response, overnight demand voids become somewhat filled; therefore, it appears that the Danish electricity production and transmission system can easily handle a 25% (and much greater) electrification of the Danish passenger vehicle fleet. This conclusion is also predicated on the assumption that the electrical supply system can continue production throughout the night. Increased demand during the night would reduce the number of hours with very low prices (zero or negative), of which there are currently 40 h or so per year, all of which occur during the night.

Local Grid

Figure 2.1 indicates the impact of demand-response tools on the main transmission grid (assuming that additional demand at off-peak times can be met). However, the Danish power system is also made up of thousands of 0.4 kW local distribution grids, and particularly during peak hours there have been concerns that these grids may not be able to withstand the addition of large amounts of EVs.

In an effort to replicate the potential impacts on a local grid, a simplified theoretical grid consisting of 100 average single family Danish houses without electric heating was created. Depending on how much additional capacity is currently in the local grid, an attempt to quantify how many of these houses could simultaneously charge an EV in the different scenarios was undertaken.

Local grid companies often do not know the exact capacity and utilization rates of their lines; however, one can get an idea of what the minimum capacity is by looking at how much demand local grids have accommodated in the past. In 2007 (as is often the case), the highest demand hour for average Danish houses without electric heating was 24 December from 16:00 to 17:00. On average, each Danish house used 1.23 kWh/h of electricity during this hour in 2007; therefore, this figure is used as a starting point regarding estimated amount of unused grid capacity.

Christmas Eve, however, is not representative of a typical day where people will be charging their vehicles after returning home from work. Therefore, as the basis hour to which additional demand will be added, the work day with the highest demand hour in houses without electrical heating in 2007 is selected, namely Wednesday 3 January from 17:00 to 18:00. In Figure 2.2 this is represented by the lighter portion.

As was the case in Figure 2.1, the darker portion in each graph represents the additional electricity demand that is forecasted to occur as the result of charging a certain number of EVs under two different scenarios. Relative to the maximum demand hour of 1.23 kWh/h per house in 2007, and indicated by the horizontal line, Figure 2.2 displays a situation where the local grid has 25% higher capacity than the maximal use of 1.23 kWh/h; that is, 153 kWh/h for 100 houses. However, given the above-mentioned uncertainty, some local grids could fall outside of these boundaries. It should also be noted that average hourly values are used here, and maximum intra-hour demand will be higher.

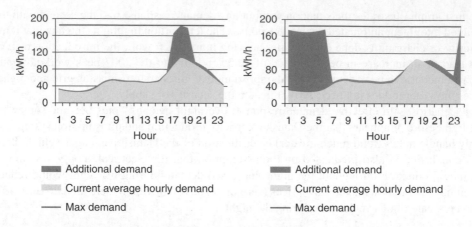

Figure 2.2   Maximum EV penetration for a Danish local grid.

Without demand response (i.e. immediate charging), the scenario in the left of Figure 2.2 reveals that with 25% unused capacity a local grid consisting of 100 average Danish houses could accommodate over 30 EVs. Meanwhile, with effective demand response and/or a fleet operator the same grid could accommodate 143 EVs. Figure 2.2 highlights the fact that a significant number of EVs can be integrated into the local grid if (a) there is at least 25% unused capacity in the distribution grid and (b) the demand-response tools are successful in spreading the charging out during the night period.

Figure 2.2 attempts to demonstrate the number of EVs that a local grid could accommodate given a fixed amount of unused capacity. Another way of approaching this involves holding the number of EVs fixed and then forecasting how much unused capacity would be required to accommodate this number of EVs.

This situation is reflected in Figure 2.3, which demonstrates the amount of unused capacity in the local grid that is required to accommodate a 25% electrification of personal vehicles within a local grid.

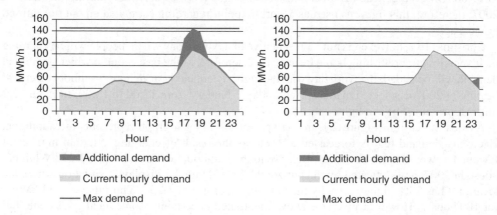

Figure 2.3   Amount of unused capacity in the local grids required to support the electrification of 25% of the Danish personal vehicle fleet.

The left side of Figure 2.3 shows a situation where 75% of EV owners begin to charge their vehicles directly after arriving home from work (immediate charging). In such a situation, if 25% of personal vehicles were to be electric, the local grid would require at least 18% unused grid capacity to cope with this additional demand. The right side of Figure 2.3, on the other hand, reveals that if effective demand-response tools are in place (market price based/fleet operator managed charging), a 25% electrification of the personal vehicle fleet would not require the utilization of any unused capacity. In fact, during the busiest hour on a work day, relative to Christmas Eve there would on average be 13% unused capacity in the local grid.

In the above discussion, the demand-response scenarios, whether they involved market-price-based or fleet-operator-managed charging, both assumed that the additional demand would be evenly spread out over the selected time periods. In practice, the signal that these tools will be responding to will most likely be the market price for electricity within a given hour. This could be problematic if the majority of users come to the conclusion that the same one or two hours are usually the cheapest and, therefore, set their vehicles to charge at this time. This would result in a situation where local grid capacity in many areas would likely be breached. As such, to be effective, any time-delayed demand-response tool will also have to be able to take into account local grid conditions, whether it be via allocating particular charging hours to customers or setting a limit on the rate of charge. The same applies to intelligent automated demand response, as an effective technology of this type should not only be able to react to market signals, but also coordinate the charging needs of other users to ensure that the collective charging is done in a more optimal fashion.

On the other hand, there will also be a degree of natural variation in the demand from the various devices used (some heat pump owners will prefer a warmer or cooler house than others, EVs will have driven different distances during the day, and will have different requirements for the next day, etc.). All of these 'natural' variations will result in varying electricity demands both in terms of quantity and time of need and, therefore, lessen the likelihood that all demand-response consumers will react precisely the same to an electricity price signal. For a more extensive discussion on how to manage congestions in the distribution grid, see Chapter 3.

2.3 Market Actors

The following section provides a brief introduction to the various actors within the Nordic electricity market.

2.3.1 Electricity Consumer: Individual Vehicle Owner

The individual vehicle owner can be an ordinary consumer in the electricity market. The EV can be regarded as a new type of consumption in line with other household consumptions, such as electrical heating, washing machines, and so on. The unique thing about the EV is that the consumption (the charging) takes place when the owner is not using the vehicle, implying that the battery allows a greater flexibility in the demand for electricity, thereby giving an opportunity to introduce demand response. The term 'demand response' refers to the fact that vehicle owners adjust their consumption according to electricity prices.

For the end-user to benefit from demand response, an interval meter is needed. These meters will be able to read the consumption per hour (or more frequently) and thereby make it possible to use price contracts with prices varying per hour (spot prices), by weekdays/weekends or day/night.

2.3.2   DSO/Grid Company

Grid companies (distribution system operators, DSOs) operate the distribution grid and are thereby responsible for the supply security through the delivery of power to the customers. In a number of countries, grid companies are obligated to ensure that meters are installed and read at all end-users. Thus, it is the local grid company who is responsible if interval meters are to be installed.

Grid companies are regulated monopolies. They must be legally and management independent from other companies (e.g. commercial companies such as generators and retailers – see to Sections 2.3.3 and 2.3.4). The Danish grid companies are benchmarked and the tariffs are controlled by the regulator.

2.3.3   Retailer

The retailer acts as a link between the power market and the consumer. The retailer communicates the consumer demand to the power market, purchases electricity from the power market and resells it to the various consumers. In the Nordic area, electricity can be bought on the Nord Pool power exchange or directly from a local generator.

Some retailers also act as a load balance responsible (LBR), and in some cases several retailers use a common LBR. The main task of the balance responsible is to make a plan of the consumption and production for the upcoming day.

In the case of imbalances (deviations from the plan) the balance responsible has to pay for imbalances to the TSO. The balance responsible can be active on the regulating power market and submit bids related to up- and down-regulation to the TSO.

2.3.4   Generator

The electricity generator produces power to the electricity system. The generators bid in with their expected power production on the Nord Pool Spot market every hour for the upcoming day. Some generators are production balance responsible (PBR) and are thus responsible for balancing the actual production according to the planned production.

2.3.5   Fleet Operators

A fleet operator is defined as an actor that operates a number of EVs in their interaction with the power system, but is not necessarily the owner of the EVs. The fleet operator can either be an independent actor or it can be the retailer and can manage the charging of the individual EVs (i.e. automate the charging and provide other services to the EVs). The fleet operator has a contractual agreement with the EV owner that describes the conditions for charging, level of control, payment, and so on.

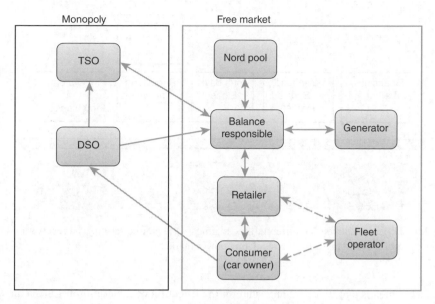

Figure 2.4   Simplified illustration of the interaction between market actors in the Nordic power system.

2.3.6   TSO

The TSO is responsible for the overall security of supply and to ensure a well-functioning electricity market by maintaining the electrical balance in the power system and by developing market rules. The Nordic TSOs own the Nord Pool power exchange.

2.3.7   Nord Pool

Nord Pool manages the Nordic power exchange, which includes the physical trade in the two markets Elspot and Elbas and the financial markets (see Section 2.4). Figure 2.4 illustrates the market actor set-up with the actors as described above. Data for the consumer's electricity consumption is recorded in the meter and sent to the DSO. After consolidation of data, they are sent to the TSO and LBR.

   The consumer enters a contract with a retailer (or with a fleet operator). Each retailer refers to a LBR. The LBR sends a plan for the next day's electricity demand to the TSO. The electricity can be bought from the power exchange Nord Pool. If the actual generation or consumption differs from the scheduled plan the difference is bought or sold from the TSO as unbalances.

2.4   Nordic Electricity Markets

The current Nordic electricity market consists of a number of specific underlying markets based on a timeline for the bidding offers. Figure 2.5 illustrates the major components of this market.

   It should be noted that the Nordic electricity markets are constantly developing. The development in the Nordic system and in Europe as a whole is towards one European cross-border

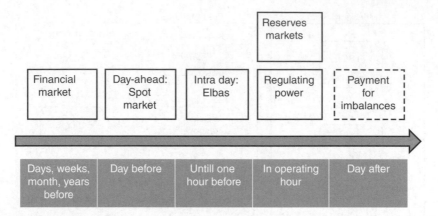

Figure 2.5   Different markets for different time regimes – the Nordic set-up. The reservation markets include reservation of resources for the regulating power market.

market. This means that unless changes are in the direction of the common European set-up, dramatic changes are not realistic in the short term. This set-up is defined in groups managed by the European regulators. All the Nordic markets are geared towards a European set-up, in large part because in many circumstances it was the Nordic market set-up that has been the basis for the broader European set-up.

2.4.1   The Spot Market and the Financial Market

The central Nordic energy market is the spot market (Nord Pool Spot) where a daily competitive auction establishes a price for each hour of the next day. The trading horizon is 12–36 h ahead and is done in the context of the next day's 24 h period. The system price and the area prices are calculated after all participants' bids have been received before gate closure at 12:00. Participants' bids consist of price and an hourly volume in a certain bidding area. Retailers bid in with expected consumption (which is often price independent) while the generators bid in with their production capacity and their associated production costs. A limited number of bid types exist; for example, a bid for a specific hour or block bids.

The price is determined as the intersection between the aggregated curves for demand and supply for each hour – taking the restriction imposed by transmission lines into account. Figure 2.6 illustrates the formation of the system price on the spot market as a price intersection between the purchase and sale of electricity.

The financial market is a market where price-securing contracts are traded. The financial markets trade futures and other derivatives that are settled against future spot prices; it is possible to enter a future contract involving, for example, 100 MW next year. Most liquidity is related to financial contracts targeting the *system price*. The system price is an artificial price that would be the result if no congestion exists. Financial contracts, options and contracts for differences (the difference between the system price and the area price in a specific price area) manage risks and are essential for the market participants in the absence of long-term physical contractual markets.

Figure 2.6 The formation of the system price for electricity on the Nord Pool Spot market (www.nordpoolspot.com).

In order to handle grid congestions, the Nordic exchange area is divided into bidding areas. The bidding areas are consistent with the geographical area of each of the TSOs. Denmark, however, is divided into two bidding areas, east and west (DK2 and DK1). The same goes for the Norwegian grid, which is usually divided into five or six bidding areas, and the Swedish grid, which is divided into four areas. Participants must make their bids according to where their production or consumption is physically located in the Nordic grid areas. In this way the transmission capacity between the different bidding areas is implicitly auctioned via the submitted bids and the spot price calculation. Thus, whenever there are grid congestions, different prices will occur in the affected price areas. The participants' bids in the bidding areas on each side of the congestion are aggregated into supply-and-demand curves. Figure 2.7 illustrates the formation of the area prices.

As illustrated in Figure 2.7, a volume corresponding to the trading capacity on the constrained connection is added as a purchase in the surplus area (export) and as a sale in the deficit area (import). In the deficit area the import will lead to a parallel shift of the supply curve, while in the surplus area the additional purchase will lead to parallel shift of the demand curve.

By increasing the price in the deficit area, the participants in this area will sell more and purchase less electricity, while in the surplus area a lower price will lead to more purchase and less sale. The area price calculation is repeated so that the transmission capacity between the high price area and the low price area is utilized to the maximum. Each TSO area may contain one or more price areas depending on where the grid congestions are from hour to hour; thus, the constellation of areas with the same price may change every hour.

In situations where the power flows are between AC bidding areas within the limits set by the TSOs (that is, when no congestions exist), the system price is the price for all price areas.

2.4.2 Elbas Market (Intraday)

Given that the time from fixing of the price, and the plans for demand and generation in the spot market to the actual delivery hours is up to 36 h, deviations can occur. Deviations can,

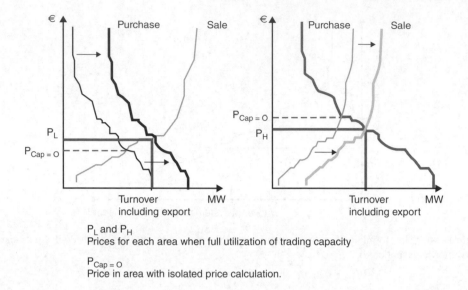

P$_L$ and P$_H$
Prices for each area when full utilization of trading capacity

P$_{Cap = 0}$
Price in area with isolated price calculation.

Figure 2.7 The formation of the area prices for electricity on the Nord Pool Spot market (www.nor dpoolspot.com).

for example, come from unforeseen changes in demand, tripping of generation or transmission lines or from incomplete prognoses for wind power generation. Such deviations can be compensated during the operational day by hourly contracts in the Elbas market up till 1 h before the operating hour. However, the liquidity in this market today is limited. Currently, the average Elbas trade in the Nordic area is only 200 MW, although this trade has seen an increasing trend [5].

2.4.3   Regulating Power Market

In order to have stability in the Nordic electricity system, different criteria must be met at all times [6]:

- The frequency of the synchronous system must be between 49.9 and 50.1 Hz.
- The time deviation of the synchronization shall be within the range [−30 s, 30 s]. Time deviation is found by integrating the frequency deviation from 50 Hz.
- The requirement in Regional Group (RG) Continental Europe that every control area has to keep its own balance.[1]
- The flow on any line must not exceed the transmission capacity at any time.

These criteria are set by the TSO associations. As of July 2009, the European TSO associations merged into ENTSO-E, covering 41 TSOs and 34 European countries. The electricity

---

[1] Western Denmark has a special requirement within the RG Continental Europe to keep its own balance in Jutland.

system is still divided into the same synchronous areas, now called RGs. Thus, Denmark is both a part of the regional Nordic group (Eastern Denmark) and the regional Continental Europe group (Western Denmark).

In the hour of operation, several types of reserves ensure stability of the system. The reserves can be grouped into automatic and manual reserves. The system criteria are initially managed by the automatic reserves.

To anticipate excessive use of automatic reserves and in order to re-establish the availability of these, regulating power is utilized. Regulating power is a manual reserve. It is defined as increased or decreased generation that can be fully activated within 15 min. Regulating power can also be demand that is increased or decreased. Activation can start at any time and the duration can vary.

In the Nordic countries there is a common regulating market managed by the TSOs with a common merit order bidding list. The balance responsibles (for demand or generation) make bids consisting of amount (megawatts) and price (euros per megawatt-hour). All bids for delivering regulating power are collected in the common Nordic NOIS-list and are sorted in a list with increasing prices for up-regulation (above spot price) and decreasing prices for down-regulation (below spot price). Taking into consideration the potential congestions in the transmission system, the TSO manages the activation of the cheapest regulating power.

In a supplement to this, Denmark also has a reservation market. Resources can receive a payment for being present in the regulating power market. A similar system exists in Norway (RKOM), which is only active during the winter period. Sweden and Finland also have a system in place where old oil power plants receive payments for being available if necessary.

There is an interaction between the spot market and the regulating power market, and the reservation market is used to attract sufficient resources to the regulating power market. For example, with high spot prices it is so attractive to produce for the spot market that a high reservation price is needed to maintain capacity for up-regulation in the regulating power market – and vice versa for low spot prices. The reservation price is established based on the amount of regulating power needed by the TSO and bids from potential suppliers. The payments for activating regulating power are passed on to the balance responsible after the day of operation.

Market Requirements

Participation in the regulating power market requires that certain elements can be fulfilled. In the case of EVs the most important preconditions are:

- minimum bid size is 10 MW;
- requirement of real-time measurement.

The minimum bid size cannot be met by the individual EV owners; so, if EVs are to participate in the regulating market, this precondition either has to be relaxed or EVs must be aggregated by a fleet operator or balance responsible. Such an adjustment is not problematic. On the other hand, the requirement for real-time measurements for the individual units will be very costly to fulfil and provides a significant barrier. It can be argued that other methods, such as statistical methods, are more relevant in relation to thousands of small units and could provide a workable solution.

2.4.4   Future Development of the Nordic Regulating Power Market

With the intense focus on climate challenges, renewable energy becomes an important tool for reducing emissions from fossil fuels. Throughout northern Europe a great deal of new offshore wind capacity will be developed in the years to come. When wind power capacity is expanded, more regulating power will likely be needed due to the limited or imperfect predictability of wind power. Togeby *et al.* [5] describe the structure of prognoses error in wind power forecasts. It was found that the largest errors occur when medium wind speeds are expected. With wind power capacity in the Nordic countries scheduled to increase greatly in the upcoming years (there is roughly 6000 MW of wind power capacity installed today, a figure that is likely to grow to approximately 15 000 MW in 2020), the need for regulating power is also expected to increase. As such, more focus will fall on the regulating power market, including the participation of new regulating power sources.

If small consumption, as in the case of EVs, were to be activated in the regulating power market, then the preconditions involving the minimum bid size and the requirement of real-time measurement should be carefully investigated.

In 2010, the Danish TSO Energinet.dk, with that purpose in mind, introduced draft thoughts of how small consumption units can be activated in the regulating power market. They suggest publishing a regulating price in the hour of operation to be used only for small consumption. The published price will be the price of the last taken bid from the bidding list (from the more traditional regulating resources). This suggestion would give the possibility to send a price signal to the EV – from both the spot market and from the regulating power market. The draft is being discussed among the Nordic TSOs but has yet to be implemented. As was noted previously, the regulating market is a common Nordic market and, therefore, consensus amongst the Nordic TSOs must be reached on any such changes.

Developments in Measurement and Price Settlement

Norway, Sweden and Finland have all announced plans for the phasing in of interval meters for most end-users, thus allowing for hourly price settlement. In Denmark meanwhile, a 'third settlement-group' has been suggested as an initial step in the same direction. The term 'initial' is used because the third settlement-group approach has less stringent requirements regarding the sending in of final data.

2.5   Electricity Price

2.5.1   Composition of End-User Price

In the context of individual EV charging it is important that the end-user price is broadcasted to the consumers and that the end-user price is composed in a way that the consumers will in practice react to the price. In the case of a fleet operator subscription, the individual EV has no use for the end-user price, as it is optimized by the fleet operator or fixed by contractual terms.

The wholesale price (the spot price) is a part of the end-user price, which also includes taxes, network payments, and so on. If we take Denmark as an example, the end-user price consists of several parts:

- *Market electricity* (20%) is the commercial part of the electricity; that is, the electricity traded on the spot market.
- *Transport* (12%) covers the costs of electricity transport from the production unit to the end-user and includes grid tariffs to the TSO and the DSO.
- *Taxes and public service obligations* (68%)
  - Various taxes (40%) are fixed proportionally to the amount of kilowatt-hours consumed by the end-user and cover CO_2 taxes, electricity taxes, distribution taxes and electrical heating taxes. For companies, the tax is much lower than it is for households.
  - Public service obligations (8%) are legal obligations paid by all consumers for subsidies for wind energy, combined heat and power and research and development.
  - VAT (20%). Finally, the costumers pay 25% VAT on the total electricity bill (thus comprising 20% of the total). Adjusting the various elements of the electricity price may result in having an electricity price that has enough variations (large differences between low prices and high prices) to make it attractive for consumers to react to price fluctuations; that is, to become flexible.

2.5.2 Fixed Tariffs for Losses

Electricity transportation losses are on the order of 1% in the transmission grid and 6% in the distribution grid. Today, the costs of losses are covered by tariffs to the TSO and the DSO. These tariffs appear in the form of a fixed value per kilowatt-hour. The tariffs do not reflect the actual price of electricity or the level of losses in the actual hour.

Ideal tariffs for losses should show the *marginal losses* at any given time. Marginal losses are, for example, the losses an extra electricity demand gives rise to. It can be noted that the marginal losses in practice can vary from −5% to 20% depending on the flow in the grid and the presence of local generation. Negative marginal losses can exist if an extra demand results in less transport; for example, in cases with export for distribution to transmission grids. Owing to the fact that losses are related to the square of the flow in distribution lines, the marginal loss is typically twice the average loss. This results in the highly varying marginal losses. With the significant amount of wind production in Nordic area coupled with the fact that large portions of the region act as electricity transit areas, it can be very difficult to determine the marginal loss values.

2.5.3 Transport and Local Congestions

Today, congestion in the distribution grids is managed by expanding or reinforcing the grid; that is, via the implementation of new cables, transformers or through a new grid layout. Each consumer has an installation with a certain maximum power rating; however, the grid is designed in a way that takes into account the low coincidence factor between different users. A user may in some situations consume power close to the maximum allowed value, but this is only seldom and not closely correlated to other users' consumption. In this way the distribution grid is efficiently designed – but it will not be able to deliver the requested power if a high fraction of the end-users utilize their allowed consumption at the same time.

2.5.4 Taxes

If we return to the example of Denmark, taxes on electricity used by households are very high. Today, all taxes are a fixed value per kilowatt-hour, and only the VAT is a percentage of the price. The Danish Energy Agency undertook a study of dynamic tariffs [7] and the Ministry of Taxation [8] studied the possibilities of using dynamic taxes in order to help the integration of wind power.

2.5.5 Future Tariff Possibilities

Dynamic tariffs could be used to balance the optimal response to congestion in the distribution grid. Ideally, this can be done as an auctioning of the available capacity (in a future with a high penetration of computers and communication) or in a simplified way with time-of-use tariffs. With dynamic tariffs the end-users may adapt their consumption to the varying electricity costs (defined as the sum of energy and transport). This could be via automatic demand response such as air-conditioners, electric heating, heat pumps or charging of EVs. The tariffs for managing congestions in the distribution system could be in the form of critical peak pricing (see Section 2.6.3) – so a high price is only used in critical situations.

2.6 Electricity Sales Products for Demand Response

The end-users' payment for electricity is defined by the contract between the end-user and the retailer. The different price models presented in this section are of relevance to all types of electricity consumption within the households, not only to that of EVs.

2.6.1 Fixed Price

With a fixed-price contract the consumers utilize electricity whenever it is needed and the price is decoupled from the actual hourly price fluctuations. This is the situation for the majority of Nordic households today. In the case of fixed-price contracts there is no need for online communication, interval metering, and so on. The fixed-price scenario does not support or provide incentives for demand response and, as such, often results in 'instant charging'.

2.6.2 Time-of-Use Products

A very simple price product could be a price with two time regimes: a low price on weekends and at night, and a higher price for other times. The time periods can be defined by the individual retailers, and the end-user is able to react to this simple tariff by, for example, utilizing a timer device.

Different retailers may offer various time-of-use (TOU) tariffs; for example, with two or three regimes and with different definitions of time frames. Owing to the fact that different retailers can choose individual definitions of the time frames, no negative impacts necessarily need to be related to a common increase of the load. However, if a large number of retailers were to choose to offer the consumers the same time periods in their TOU contract, then there could be a problem in the local grid with a large number of customers activating their consumption at the same time, namely the first hour of the cheapest period. Thus, a variation in the retailers' definitions of time frames is preferable.

Examination of the spot price indicates that TOU products may prove to be efficient. If one takes, for example, the average Danish spot price on weekends and at night (from 22:00 to 07:00), it is 0.031 €/kWh, while on workdays it is 0.045 €/kWh – a price difference of 45%.[2]

The TOU tariff defines periods of high and low prices independent of the actual prices. In this way, a TOU tariff averages out many of the extreme price variations. It is a simple tool that can provide better utilization of the power system.

2.6.3   Critical Peak Pricing

The TOU tariff may lead to a general shift in demand – but it is not specifically targeted at the infrequent extreme prices. A retailer may, therefore, choose to use a critical peak price (CPP) in order to target unusually high or low prices. A CPP can be a fixed price combined with a high and a low price that can be activated; that is, with a day's notice. A price product could thus be 0.040 €/kWh for all hours except for 50 low-price hours per year and 50 high-price hours per year – all announced the day before.

As was the case with the TOU tariff described above, it is important to note that each individual retailer can choose its own definition for the criteria of the CPPs. In this way, a situation can be prevented where all costumers start consuming directly after the critical peak hour, and thereby creating a new peak. This product demands computerization and a higher level of communication compared with the TOU tariff. Nevertheless, the CPP will enable some degree of demand response around the hours where the high peak price is scheduled.

2.6.4   Spot Price

The most detailed/advanced contracts include price signals for every hour and imply broadcasting of the spot price. With a spot price (or a day-ahead price) the end-user pays the actual price for electricity in each hour and an interval meter keeps track of the actual hourly consumption. The price is for example published at 13:00–14:00 the day before, for each hour of the day. The end-user in this scenario is motivated to move the electricity consumption from expensive hours to less expensive hours. Typical spot prices are on the order of 0.04 €/kWh, but can be zero or even negative. High prices around 0.20 €/kWh also occur.

The end-user can react to the general pattern in the prices (e.g. lower at nights) or can react to the actual values. In order to react to actual values, automation solutions should be used, for example, as demonstrated in the projects with spot prices for households with electric heating

---

[2]Data from West Denmark (DK1), 1 January 2002 to 19 January 2009.

[9]. An alternative is to inform the end-user when extreme prices occur. If this takes place a few times per year then the end-user can manually adjust the consumption.

A precondition for obtaining some level of demand response in the spot price scenario is a communication system that enables broadcasting of the hourly spot prices for the following day to the consumer.

Remote Control of Demand

With a spot price, with or without financial contracts, the end-user is in charge of the demand response (e.g. timing of the charging of the EV). End-users can also choose to give the task of optimizing the charging to the retailer or a fleet operator. The retailer may offer a simple contract; for example, with a fixed price – but with a discount compared with the traditional fixed price. Different retailers and fleet operators may offer different types of user interfaces. One retailer or fleet operator may decide the EV charging independent of user needs, while others may give room for some or full user influence. An advantage of this set-up could be that the retailer and fleet operator may reduce the cost of the automation and control systems due to a large volume. However, this set-up demands a two-way communication system where the retailer or fleet operator has knowledge of the state of the individual EV, such as driving patterns, charging preferences, state of charge of the battery, and so on.

2.6.5 Future Contract Possibilities Including Regulating Power Market

The retailer can also include regulating power in the pricing in order to take advantage of the high variation in prices in the regulating power market. For end-users with demand that can be controlled, regulating power can reduce the electricity cost (e.g. by dispatching demand to periods with low prices). To be active in the regulating power market the electricity demand can be remotely controlled, or the end-user can receive a dedicated price signal indicating the need for regulating power.

2.7 EVs in Different Markets

This section introduces three possible interactions between the EV and the electricity market, starting with the simplest model for a contract structure between the vehicle owner and the market. The three different charging strategies considered are:

1. *Immediate charging.* The vehicle charges whenever the consumer needs it to, regardless of the electricity price.
2. *Time-delayed charging.* The vehicle charges in predefined time periods; for example, during the night. The charging is controlled by a simple timer.
3. *Market-price-based/fleet-operator-managed charging.* The vehicle charges in the cheapest hours; for example, the three cheapest hours until 07:00 the next morning based on received price signals. An advanced version of this scenario involves introducing a fleet operator to optimize the charging strategies for an aggregated pool of EVs.

All three charging scenarios will fit into all three contract structures except for minor changes in the market rules under the third contract structure, which will be described below.

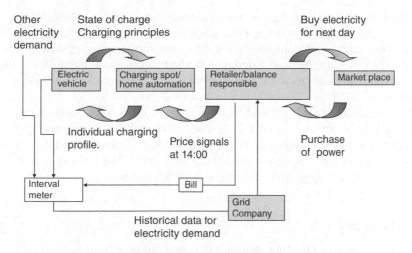

Figure 2.8   Communication with contract structure 1.

2.7.1   Contract Structure 1: The Current Spot Market

Under contract structure 1, EVs can interact with the current spot market without any changes in market rules. The EV owner with a conventional manual offline meter must charge at a fixed price as they cannot make use of intelligent charging. With time-fluctuating prices in place the end-user can decide to react to these in a simple way or in a more advanced way.

The EV owner with an interval meter can decide to buy electricity via a type of contract with varying prices. Assuming the introduction of other electricity sales products, this could be a simple TOU contract (e.g. with low prices at nights and weekends) or a CPP contract (e.g. with occasional high and low prices).

More intelligence can be obtained with an interval meter combined with a home automation device reacting to hourly signals as spot prices. This combination gives the opportunity for the EV owner to utilize the fluctuations in the electricity prices and thus obtain more favourable prices. As discussed in Section 2.5, the composition of the end-user price can in the future include time variation of tariffs for transport of electricity and taxes to give more time variation in the price.

Figure 2.8 illustrates the interaction and communication between the end-user (EV) and the retailer in the electricity market using only spot prices announced every day at 14:00.

Under this scenario, prices are broadcasted from the retailer and received by the household. The only information going from the end-user to the retailer is the historical consumption data. Based on these data, the retailer plans for the next day.

Electricity is Bought Based on Historical Data

As illustrated in Figure 2.8, the electricity demand of the household is registered by an interval meter that delivers historical data about the hourly electricity demand via the grid company to the retailer. These data are based on the EV consumption and other electricity consumptions of the household. A home automation device connected to the EV could deliver information on its state of charge and its predefined charging principles to the charging spot.

With a large number of consumers with EVs, the retailer can forecast the next day's aggregated EV electricity demand based on historical consumption data. This is the current procedure applied to most end-users concerning electricity consumption, including small and large end-users.

Based on historical data for demand for all costumers, the retailer communicates the EV's demand (and other electricity demand in the household) to the electricity market for the next day. When the spot market closes and the prices for the next 24 h are fixed, the retailer delivers signals to the EV charging spot; for example, price signals. The EV can now, based on its individual charging strategy, charge according to known prices.

Charging Strategies

As mentioned above, the charging strategies that were introduced in Section 2.2 (immediate charging and market-based charging), as well as time-delayed charging, can function in the current spot market with the introduction of an interval meter. The EVs can interact with the current spot market without any changes in market rules via immediate charging, time-delayed charging or market-price-based/fleet-operator-managed charging. The fleet operator scenario is further treated in Section 2.7.3.

The time-delayed charging can be based on charging needs and historical prices. One EV owner may, for example, set the charging to start at midnight and another owner at 22:00. This will avoid charging at 17:00–18:00 – the typical time for high prices – and will, therefore, charge in many of the hours where zero or negative prices can be realized. With this strategy the end-user may analyse the prices once a year and, therefore, not need the daily information about the next day's prices. This simple strategy may be optimal because of the low investment cost related to the automation. However, the strategy will not always target the very lowest prices.

With some kind of home automation in place the EV owner may choose a more advanced strategy. If the retailer communicates the electricity prices for the upcoming 24 h, the EV owner can choose hours to charge based on predefined needs (e.g. it has to be fully charged by 7 a.m.). The home automation device could automatically choose the cheapest hours within the defined charging period. The strategy can be designed so that a plan is defined by the home automation device based on the consumer's predefined needs each time the EV is connected to the charger. Every time the EV is disconnected and connected again, the charging automation will automatically create a new charging strategy based on known electricity prices.

2.7.2   Contract Structure 2: The Spot Market and Regulating Power Market

Under contract structure 1 the EV is reacting to the spot prices only. Relative to a fixed-price contract, this makes it possible to reduce the costs of charging. However, the costs may be reduced further if the EVs are exposed to prices reflecting regulating power prices. This is because the price variation for regulating power is larger than for spot prices, and the battery capacity can be utilized to a greater extent. Regulating power is more or less defined as consumption that can be activated within 15 min.

This contract structure requires changes to the current rules. Three challenges are associated with handling EVs in the Nordic regulating power market:

1. It is currently required that suppliers of regulating power must have online measurements. This would be expensive for a fleet of thousands of EVs. In the following, it is assumed that this requirement is relaxed.
2. The current requirement of a minimum bid size of 10 MW prevents EVs from participating separately in the regulating power market. Participation would require an aggregation of thousands of EVs to reach the 10 MW requirement.
3. The activating or stopping of charging of EVs due to regulating power will change the expected charging profiles/strategies and thereby alter the reliability of the historical data that the demand bids on the spot market are based upon.

For the EVs to act on the regulating market there needs to be a communication between the retailer and the EV. This communication can be a one-way price signal where the price signal can motivate the EV to alter its charging plan. Charging can be activated or stopped according to price signals from the regulating market and the retailer can predict the effect based on historical data. This set-up is simple and can deliver certain accuracy in the prediction. A more advanced solution is described below in Section 2.7.3.

Figure 2.9 illustrates the interaction and communication between the end-user and key actors in the electricity market using the spot prices and the regulating power market. Compared with contract structure 1, price signals are sent to the charging spot/EV at 14:00 and when regulating power bids dictate the need for additional price signals. Only local control of the EVs is considered in this contract structure. The market place is now both the Nord Pool Spot market and the Nordic regulating power market, but the system requires changes in the current market rules.

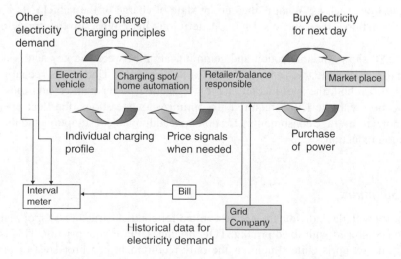

Figure 2.9 Communication with contract structure 2.

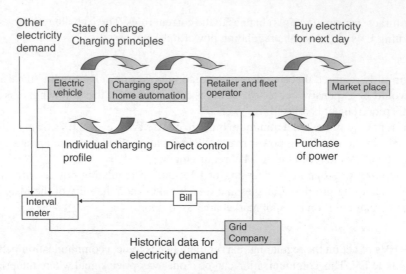

Figure 2.10   Communication with contract structure 3.

2.7.3   Contract Structure 3: EVs Controlled by a Fleet Operator

The requirement for predictable results in the regulating power market can be met by introducing more advanced communication equipment with online measurement. In order to be able to prove a given output in the regulating market, more information is needed. Within a centralized control system the information flow can be collected, including the status of the EVs, by a fleet operator.

The individual vehicle will need a price signal, whether it comes directly from the electricity retailer or goes through a fleet operator. In the case of using a fleet operator, the fleet operator will need historical or statistical data on all the EVs under its management and be able to answer questions such as: What is the current state of charge of the vehicle? Is it currently charging? What does the vehicle's driving pattern look like? What are the pre-definitions of the vehicle?

Figure 2.10 shows the interaction and communication between the key actors in the spot market and regulating market with the use of a fleet operator. Under this scenario the fleet operator bids on both the spot market and the regulating power market, and thus real-time communication with the EV is required. To optimize its operations, the fleet operator will need to know the average consumption, the potential additional consumption and the potential consumption reduction.

2.7.4   Summary

Local control with the individual EV is best suited for contract structure 1 (spot market only) and can be combined with fixed pricing, TOU pricing, CPP or spot pricing. This is possible today and the set-up is quite simple, as the only requirement is a broadcast of prices once a day.

Central control through a fleet operator is more relevant to contract structure 3, where EVs are operating on the regulating market as well. Contract structures 2 and 3 could imply more attractive prices for the consumer (the EV owner), but also require more advanced communication (real time) and this is not possible today because of the framework requirements from TSOs.

References

[1] Ea Energy Analyses (2009) CO_2 emissions from passenger vehicles. Ea Energianalyse A/S, September.

[2] Danmarks Statistik (2009) Nøgletal for Transport 2008. Danmarks Statistik, February 2009.

[3] Dansk Energi (2009) Elforsyningens tariffer & elpriser pr. 1. januar 2009, May.

[4] Dansk Energi (2009) Vejledning for tilslutning af ladestandere i lavspændingsnettet, November.

[5] M. Togeby, J. Werling, J. Hethey et al. (2009): Bedre integration af vind – analyse af elpatronloven, treledstariffen for mindre kraftvarmeanlæg, afgifter og andre væsentlige rammebetingelser. Ea Energianalyse og Risø DTU, Copenhagen, June.

[6] Nordel Demand Response Group (2006) Enhancement of demand response.

[7] Danish Energy Agency (2010) Redgørelse om mulighedrne for virkningerne af dynamiske tariffer for elektricitet, June.

[8] Skatteministeriet (2010): Redegørelse om muligheder for og virkninger af ændrede afgifter på elektricitet med særlig henblik på bedre integration af vedvarende energi dynamiske afgifter. May, 2010.

[9] DI-Energibranchen (2009) Pristølsomt elforbrug i husholdninger. DI – Energibranchen, SYDENERGI a.m.b.a., SEAS/NVE a.m.b.a., Siemens A/S, Danfoss A/S, Ea Energianalyse A/S, August.

3

Electric Vehicles in Future Market Models[*]

Charlotte Søndergren,[1] Christian Bang,[2] Camilla Hay[2] and Mikael Togeby[2]

[1] *Danish Energy Association, Lyngby, Denmark*
[2] *Ea Energy Analyses A/S, Copenhagen, Denmark*

3.1 Introduction

Chapter 2 described the current Nordic electricity market and addressed the challenges related to a large-scale introduction of electric vehicles (EVs). This chapter looks further down the road and discusses future market models that can be utilized to incorporate EVs. Sections 3.3 and 3.4 introduce different possibilities for alternative market models that could be relevant for a large-scale introduction of EVs and solve some of the challenges that are connected to the current market design. To a large extent, the point of departure for the discussion is the existing Nordic market model.

In Section 3.5, different methods to handle the challenges associated with congestions in the distribution grid are described.

3.2 Overview

This section discusses various potential markets and actors that could be relevant for EV users allowing EVs to provide regulating power and automatic reserves and dealing with potential problems that large-scale EV integration may pose. It also deals with how to handle congestions in the distribution grid. A number of different types of approaches where both the wholesale market and the grid state are used to set the dispatch are listed.

---

[*] This chapter presents the findings from report 2.3 of the Edison project, "Electric vehicles in future market models," and the FlexPower project (www.flexpower.dk).

Grid Integration of Electric Vehicles in Open Electricity Markets, First Edition. Edited by Qiuwei Wu.
© 2013 John Wiley & Sons, Ltd. Published 2013 by John Wiley & Sons, Ltd.

3.2.1   Spot Market

The easiest market for EVs to participate in is the spot market, as it simply requires an EV user to enter into an agreement with a load balance responsible (LBR) and have a meter that can measure on an hourly basis.

However, if a large number of EVs participate in the spot market, congestions in the distribution grid could arise, congestions that the current market structure cannot deal with. The so-called lemming effect can arise, as all EVs can choose to charge in the same (and cheap) hour, if no (price) signal is provided from the distribution grid to avoid this. As such, the utilization of a fleet operator and/or various market tools, such as locational pricing or congestion tariffs, could be the solution.

3.2.2   Regulating Power Market

If EVs are to gain access to the regulating power market, then the existing additional requirement of real-time measurement is problematic (i.e. expensive) and will either have to be relaxed or addressed. If the requirement is relaxed for smaller units, concepts such as Flex-Power (described in Section 3.3.4) will become viable. If the bid requirement is also relaxed, a 'self-regulating' concept will become relevant, as it places the responsibility for anticipating the demand response in the hands of the transmission system operator (TSO) and, thus, does not require the retailers to send bids for delivering regulating power. As is the case with EV participation in the spot market, congestions in the distribution grid also require a fleet operator and/or new various market tools.

3.2.3   Automatic Reserves

In order to participate in the market for automatic reserves, EV owners will either have to enter into an agreement where they receive compensation in return for allowing their vehicle to stop charging when the frequency drops (similar to the demonstration project with demand as frequency reserves as described in Section 3.3.6) or enter into an agreement with a fleet operator who will be able to provide automatic reserves by pooling EV users. In the first concept, a device that automatically cuts out when the frequency drops below a predetermined set point is required. The fleet operator concept, meanwhile, requires a technology to allow for the fleet operator to control the charging of the vehicles, and in this regard could provide frequency control in either direction (similar to the Delaware project described in Section 3.3.7).

3.2.4   Congestions in the Distribution Grid

Congestions in the distribution grid can, in principle, be handled in two ways: either by direct control or by a market-price approach (indirect control). How to handle congestions in the distribution grid depends on which market the EVs participate in. If the same accuracy is to be obtained, the more markets the EVs participate in, the more complicated the solutions will be. Involvement in the continuous regulating power market will also require continuously updated price signals corresponding to the congestions in the distribution grid if, for example, dynamic tariffs are considered.

Congestions in the distribution grid can also be handled by allowing direct control from the grid company. This solution is less accurate, but is simpler. Alternatively, the distribution grid has to be reinforced. Comparing the cost of these alternatives will give an indication of what instrument to use in which case.

A mixed approached to handle system balance and distribution grid congestion is described. Suited to a virtual power plant (VPP) in the Edison project, it lives up to minimum requirements in such a way that the approach fits the current system with no need for changes. It considers EVs both in the spot and in regulating power market, and is relatively socio-economic and simple. The method can be somewhat simplified if congestions in the distribution grid are handled through direct control.

3.2.5   Role of the Distribution System Operator

Management of congestion in the distribution grid and taking into account end-user reaction to price signals with tools other than grid reinforcements are a new and untested area. All methods require more knowledge of the distribution grid than is available today.

In addition, new methods to handle congestions in the distribution grid and to predict the reaction of the consumers have to be developed. In particular, mapping of congestions in the distribution grid, geographical differentiation of end-users, pricing through a compensation mechanism and developing new tools for including price-dependent demand are obstacles to be overcome before distribution grid congestions can be handled in a new manner.

3.3   Alternative Markets for Regulating Power and Reserves for EV Integration

In order to facilitate greater integration of EVs, alternative electricity market models and aspects of the market are considered. Potential markets that could be relevant for EVs are the regulating power market and the markets for automatic reserves. Sections 3.3.1 and 3.3.2 briefly describe the regulating power market and the market for automatic reserves. Sections 3.3.3–3.3.7 describe concrete proposals of how small-scale consumption units such as EVs could potentially be incorporated into the Nordic regulating power market and what potential changes to the market would be necessary in order to make the market relevant for EVs.

3.3.1   The Regulating Power Market

As electricity production and consumption always have to be in equilibrium, deviations in the final operating hour are left for the TSO to balance. In the Nordic countries this is done via the regulating power market. As renewable energy will become an increasingly important tool for reducing emissions from fossil fuels, production will become more intermittent (solar, wind, etc.); therefore, it is anticipated that the need for regulating power will increase.

In the Nordic countries there is a common regulating power market managed by the TSOs with a common merit order bidding list. The balance responsibles (for load or production)

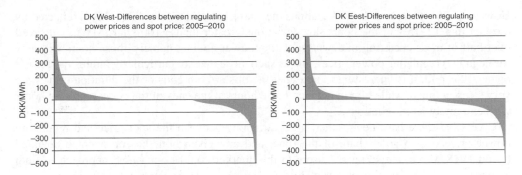

Figure 3.1 Historical differences (2005–2010) between regulating power prices and spot prices: (a) DK West; (b) DK East.

make bids consisting of amount (MW) and price (€/MWh). In Denmark, the minimum bid size is 10 MW and the maximum is 50 MW. All bids for delivering regulating power are collected in the common Nordic Operational Information System (NOIS) list. These bids are sorted in a list with increasing prices for up-regulation (above spot price) and decreasing prices for down-regulation (below spot price). Bids can be submitted, adjusted or removed until 45 min before the operational hour. Taking into consideration the potential congestions in the transmission system, the TSO manages the activation of the cheapest regulating power via this NOIS list. Owing to the fact that regulating power is specialized in nature (it must be fully activated within 15 min and the duration may vary), it demands a price premium with respect to the spot market. It is this greater price volatility that makes the regulating power market particularly interesting for EV users. With access to this market, EV users will be able to take advantage of these price variations and charge their vehicles when prices are low and avoid charging when prices are high. Figure 3.1 displays a duration curve of the hourly differences between the regulating power prices and the spot price for West and East Denmark from 2005 to 2010. The average spot price over the period was 309 DKK/MWh in DK West and 325 DKK/MWh in DK East (€1 ≈ 7.44 DKK). For illustrative purposes the vertical axis has been limited to ±500 DKK/MWh; however, maximum and minimum values greatly exceeded these values.

Figure 3.1 and Table 3.1 highlight a number of relevant aspects. First, in roughly 25–33% of the hours the regulating power price has deviated less than ±1 DKK/MWh from the spot price.

Table 3.1 Historical differences between regulating power prices and spot price in West and East Denmark from 2005 to mid August 2010.

	DK West	DK East
Average spot price (DKK/MWh)	309	325
Average absolute difference (DKK/MWh)	65	65
Minimum difference (DKK/MWh)	−6566	−10 136
Maximum difference (DKK/MWh)	7034	14 712
Hours with a difference less than ±1 DKK/MWh (%)	33	25

However, when there have been deviations they have been rather sizeable. This is reflected by the fact that 12–15% of the hours saw deviations greater than ±100 DKK/MWh, with well over half of these occurring in down-regulating power hours – namely, those hours with low (or negative) regulating power prices, and as such those that are particularly interesting for EV owners. Second, it is interesting to note that the tips at both ends of the duration curves are very steep; as such, while rare in number, those hours at the ends of the curve represent hours with extremely large variations.

Given that there have been rather large variations between the spot price and regulating power price over a good share of the time, and any given hour has on average deviated by 66 DKK/MWh, attracting EV users to this market could be a viable approach to both encouraging EV penetration and providing a necessary balancing service.

3.3.2   Market for Automatic Reserves

While the regulating power market is utilized to deal with the larger deviations within the operational hour, smaller deviations are dealt with by the fast-acting automatic reserves. There are different types of automatic reserves and, depending on the type, the providers of those automatic reservesy can receive both a reserve payment and an energy payment if activated. As the term indicates, they are activated automatically and are done so to ensure that the system frequency is kept within 49.9–50.1 Hz. They are expensive and, therefore, are potentially relevant for EVs.

3.3.3   Proposal from the Danish TSO on Self-Regulation

In order to increase the opportunities for small-scale consumption units to participate in the regulating power market, the Danish TSO Energinet.dk published a paper in 2010 entitled 'Udvikling af rammer for regulerkraft – indpasning af mindre forbrugsenheder og andre mindre enheder i regulerkraft-markedet' (Energinet.dk, 2010).

Energinet.dk's proposal is based on the understanding that increasing amounts of wind power will require more regulating power and/or large investments in new reserves and expansion of the transmission network. Energinet.dk deems that the costs associated with large transmission expansions are much higher than those related to increasing the amount of regulating power. As part of its obligation to secure the best possible conditions for competition in the market, Energinet.dk therefore investigates opportunities for expanding the framework for the regulating power market so that electricity demand and smaller units can be more active in the regulating market. Energinet.dk (2010) describes the proposal regarding how the framework for regulating power in particular can be adjusted. Energinet.dk envisions that these proposals shall be implemented in the upcoming years as part of a larger move towards a 'smart grid', which is expected to develop over the next 10–20 years.

When discussing possible alterations to the regulating power market framework, it is important to keep in mind that, today, the regulating power market is common for the Nordic countries. Thus, any suggested changes to the Danish market have to be seen in this larger context, a context that is expected to grow even larger as the Nordic market becomes further integrated with the European market. That being said, Energinet.dk does foresee some potential changes. First, it is likely that timeframes for regulating power will become shorter, thus increasing the flexibility of the market and moving the price setting of regulating power closer to the actual operating time. Another change could be the publication of the regulating power price during the hour of operation (as opposed to now, where this price is published an hour after the operational hour). In particular, the publication of the regulating power price within the hour could allow for greater EV participation in the market, a possibility that will be discussed in greater extent below.

Generally speaking, Energinet.dk outlines two methods of incorporating demand into the regulating power market:

1. Participation in the regulating power market under the current rules with bids to a common merit order list.
2. Self-regulation.

Participation in the Regulating Power Market Under Current Rules

As long as the bids fulfil the existing criteria (minimum bid size, real-time measurement, ability to reactivate, etc.), the existing market rules do allow for demand to participate in the regulating power market. However, the criteria can be quite problematic for smaller units. For example, the above-mentioned minimum 10 MW size restricts individual unit participation such as EVs from participating in the regulating power market. To work around this problem, on the production side there is already a tradition of pooling assets by the balance responsibles, thus enabling smaller units to participate in the regulating power market. Energinet.dk expects, and will support, a similar development on the demand side (Energinet.dk, 2010).

Self-Regulation

A more novel approach suggested by Energinet.dk is what it refers to as self-regulation. The concept proposes to give small-scale consumption units the possibility to react on a regulating power price. It is intended that the regulating power price will be published within the actual operating hour as the price of the most recently activated bid on the common Nordic market. By using this model, the consumption units will have the opportunity to earn profits in the regulating power market by self-regulation (via the balance settlement). To benefit from this system, end-users must have interval meters installed; however, they do not need real-time measurement.

This self-regulation can also be broken down into two main categories. In the first, the balance responsible has remote control of some of the end-users' consumption units according to a predefined agreement; for example the charging of EVs or activation of heat pumps. In exchange for relinquishing day-to-day control of these devices, end-users will in some way be compensated with lower electricity prices. Based on these agreements, the balance responsible can utilize these units to participate in the regulating power market and generate additional revenue.

In the other category, the LBR settles the electricity usage according to the regulating power price. The key aspect under this approach is that no bid from smaller demands is submitted to Energinet.dk. The end-user (e.g. through a trader, fleet operator) simply receives a price signal and acts accordingly. As above, the price signal would be based on the latest activated regulating power bid from the larger power plant (plus an additional price). However, the regulating price sent out in the hour will not necessarily be the final regulating power price for that hour, as this can first be determined after the fact when bottlenecks between various areas have been factored in. The risk caused by the potential difference between the price signal and actual regulating power price is one that in Energinet.dk (2010) is stated should be orientated by the commercial actors in the market as part of their business model. Under this model, Energinet.dk will have the task of predicting the amount of power activated by self-regulating. Data for self-regulating would be compiled and utilized by Energinet.dk in its prognosis models; that is, determine how much self-regulation will occur given a certain price signal. In Energinet.dk (2010) it is pointed out that even under the current system it is possible for LBRs to undertake self-regulation based on estimates of what the regulating power price is likely to be. This, of course, involves a risk, a risk that would be decreased if the regulating power price were publicized during the operational hour.

Automatic Reserves

While the Energinet.dk proposal mainly focuses on manual reserves and the regulating power market, it also mentions that if they can display the required flexibility and can react to the signals that are used to activate reserves, then consumption devices can also participate in the market for automatic reserves. Section 3.3.7 describes a market in the American state of Delaware where EVs are participating in such a market.

3.3.4   FlexPower

Supported by Energinet.dk, the FlexPower project investigates the potential for using demand as a stable and low cost resource for regulating power. As a starting point, current regulating power will exist and function as today, and contribute with the main volume of the regulation market. The idea is to develop a price signal that changes every 5 min and is broadcasted to all end-users interested in participating. The response should be voluntary and the price signal acts as the final settlement. The end-users that could be interested in participating in this system would have some electrical appliances that are suitable for control, with EVs being a prime example. End-users would typically install some automated control system that could receive the broadcasted price and realize the relevant control.

Within the project, a simple and efficient market will be designed and tested. The concept is similar to the above-described self-regulation variation, in that the market will make use of one-way price signals to activate electricity demand and small-scale generation as regulating power. The price will be high when up-regulation is required, and the price will be low when down-regulation is required. The price shall correspond to the latest bid activated in the traditional regulating power market. The major difference between the FlexPower project and the self-regulation proposal above is that in FlexPower it is the balance responsibles that will predict the aggregated impact based on historical data, and based on this send bids to Energinet.dk.

What makes the FlexPower project particularly interesting is that it can develop in either of the two directions highlighted by Energinet.dk: by fitting into the existing market or via a form of self-regulation. Under the first format, a balance responsible is utilized to pool items and make bids that comply with the existing 10 MW bid size. When bids are activated by the TSO, the balance responsible then sends out a price signal it deems to be sufficient to activate its small end-users to deliver the amount of regulation required by the TSO. These individual units are then monitored to ensure delivery of bids, thus again complying with existing rules.

Under the self-regulation variation, the price signal sent by the TSO is sent directly to the end-user. There is no activation of bids as the end-user simply reacts to the price signal and the TSO is responsible for forecasting what this reaction will be. The major difference between the two forms is that in the former the responsibility for predicting the demand response (and, therefore, the responsibility if actual demand deviates from the bid) falls to the balance responsible, while in the latter it falls to Energinet.dk. There are valid arguments for placing this responsibility in the hands of either party. Energinet.dk has access to more data regarding the system as a whole; therefore, it could be argued that it is better suited to undertake the prognosis and prediction associated with determining the aggregated demand response (e.g. from customers from different balance responsibles). On the other hand, in an increasingly more liberalized electricity market, it could also be argued that this task should be undertaken by private market actors, and the risks associated with forecasting inaccuracies should be incorporated into their business models. The self-regulation variation is also complicated by the fact that a balance-responsible agent could be responsible for unbalances (i.e. deviations between the amount of electricity it purchased on the spot market and the amount that was actually used by its customers) that are caused by price signals sent by Energinet.dk, thus signals that the balance responsible has no control over.

The current FlexPower set-up utilizing a balance responsible that bids in on the regulating power market is displayed in Figure 3.2. The various arrows indicate monetary flows, signals and billing, data flows, and bidding.

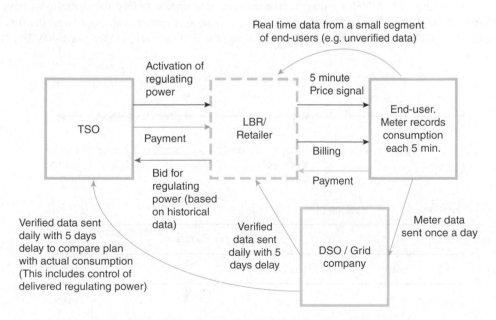

Figure 3.2   Overview of the FlexPower concept.

Figure 3.2 displays a FlexPower set-up that, with one exception, could fit within the existing market rules; this exception refers to the current requirement that all regulating power units be measured in real time. The FlexPower concept loosens this restriction, and instead allows for the end-user data to be sent to the distribution system operator (DSO) once a day. After having purchased the anticipated amount of required electricity on the spot market, the LBR submits a bid(s) to the regulating power market based on the historical price data available at its disposal. Along with all other regulating power bids, it is sorted on the Nordic merit order list, the NOIS list. When the TSO activates the bid it informs the LBR as it would with any other regulating power bid. Based on historical data and the newest information available to it, the LBR determines what price signal shall be sent to its FlexPower end-users in order to achieve the desired amount of regulation.

The end-user, most likely via some form of automation, then adjusts their consumption in accordance with the price signal. If the TSO has activated up-regulation, then a price higher than the spot price will be sent to the end-user; and if down-regulation, a price lower than the spot price will be received. Any response by the end-user is voluntary and the consumption is metered in 5 min intervals corresponding to the received price signals. This data is stored locally and sent to the DSO once a day. Taking into consideration a delay of up to 5 days to ensure quality control, this data is then forwarded to the LBR and TSO. Though, it is also possible that unverified data will be sent earlier to both the LBR and TSO for use in data analysis.

Based on this data, the TSO can then settle any unbalances between the amount of regulation promised and the amount actually delivered.

Figure 3.3 displays a timeline of the events under FlexPower in an example where a bid is activated by the TSO at 08:22, and this price signal is sent from the LBR to the end-user at 08:25. Bids can be updated continuously, until three-quarters of an hour before the operating hour. Updating bids at 07:15 would be the last possible update before the operational hour starting 8:00. Figure 3.3 highlights the fact that a number of hours can pass from the time when the spot market has settled (13:00) and when the first activation takes place (08:22). It

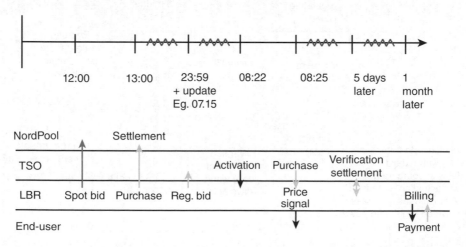

Figure 3.3   Timeline of the FlexPower concept.

is practical, therefore, for the LBR that regulating bids can be updated up until 45 min before the operating hour. In the above example, the latest update for the 08:22 activated bid could thereby have been sent in at 07:15.

FlexPower Summary

The advantage of the FlexPower concept described above is its simplicity for the end-user and Energinet.dk, and its ability to fit into the existing regulating power framework and rule set. An EV user participating in FlexPower simply predefines the charging needs and the automation will respond to price signals accordingly. Meanwhile, for Energinet.dk, FlexPower merely represents an additional regulating power bid that can be activated in the same way that existing bids are activated. It is possible that FlexPower activations will be less accurate than traditional activations, thus potentially resulting in an additional bid having to be activated. This is an area that will be investigated within the FlexPower project.

The complexity in the above concept lies with the balance responsible, as they must generate price curves and address a number of financial and forecast related risks. A key part of the FlexPower project involves the design and testing of the various prognosis and modelling tools and technologies that will be utilized by the balance responsible and the end-users.

3.3.5 Other Potential Alterations to Regulating Power Market

In addition to the Energinet.dk and FlexPower proposals outlined above, other changes to the current market rules have also been suggested. In December 2008 Nordel released a report entitled 'Harmonisation of balance regulation in the Nordic countries' (Nordel, 2008) that provided an analysis of differences between the Nordic regulating power markets and suggestions for how they can be further harmonized. In addition, the report also referred to future potential changes.

While the minimum bid size in Denmark is currently 10 MW, Energinet.dk can activate part of a bid after agreement with the bidder. This could be particularly applicable for EVs, as bids based on a number of EV users are well suited to reacting on partial bid sizes. In addition, Nordel also opened the door to smaller bid sizes in the future, another aspect that suits small end-users such as EVs. Until today, bids in the Nordic countries have to a large extent been handled manually. Particularly as more of the Nordic countries move to automated communications of bids (Denmark did so in 2008, Sweden has introduced it, and Norway has a project to address automation), this will allow for smaller bid sizes, something that the report indicates would help to promote demand-side bidding (Nordel, 2008).

Another topic that the above-mentioned report touched upon was the potential for other types of bids being included in the common Nordic lists of regulating bids, the NOIS list. Faster responding bids, for example, could be earmarked on the bid list and utilized as special regulation. The idea is that these faster bids should not receive any preferential treatment on the NOIS list when used for normal balance regulation, but could be utilized in special situations and, thus, taken out of order. Related to this, the report also suggested that slower bids could be placed on the NOIS list and also utilized as a special regulation, although in this case only when all normal bids have been used.

3.3.6 Demand as Frequency-Controlled Reserves

The frequency has to be kept between 49.9 and 50.1 Hz to ensure balance between production and demand in the Nordic electricity system. When a breakdown somewhere in the system occurs, the frequency will drop and the balance must quickly be re-established by using reserves. Today, these reserves are provided mainly by generation-side resources, including extra capacity of generators and interconnection lines with reservation that could otherwise be used for transactions in the electricity market. It is worth noting that the number of minutes per month that the frequency in the Nordic synchronous system is outside the desired ± 100 mHz band is over 10 times greater than the stated goal; therefore, there is a focus on improving the quality of frequency going forward (E-bridge, 2011).

In the future, frequency reserves could also be provided by demand units reducing their consumption. This can be provided by using frequency-controlled demands. In particular, thermostatically controlled loads such as heaters and refrigerators have cyclic on/off characteristics with considerable volume, which makes them ideal for use as frequency-controlled reserves. Domestic demands with small electricity consumption, such as EVs, can also provide reserves.

A demonstration project entitled 'Electricity demand as frequency controlled reserve (DFR)' (Demonstration Project, 2011), funded by Energinet.dk, was carried out on Bornholm in 2011 and 2012. In this project, automatic control devices were installed at approximately 200 electricity customers, involving consumption units that were suitable for automatic deactivation or reduced consumption for short periods (typically between a few seconds and 1 min). Four types of consumption participated in the demonstration project:

- Fifty bottle coolers placed in shops with a built-in automation that increases the temperature set-point of the cooler by up to 2 °C according to the current frequency.
- Fifty Danfoss thermostat control units in houses with electrical heating that adjust the temperature set-point according to the current frequency.
- Thirty Electronic Housekeepers, which are wireless home-automation units that can control selected consumption units, such as refrigerators, freezers, electrical heating, heat pumps, and so on. The control units turn off the connected devices during frequency drops.
- Fifty intelligent control units suited for larger consumptions in companies, institutions, and so on, with suitable consumption for frequency control.

The demand response obtained by the participating consumers was recorded and analysed during the project in order to evaluate the performance of the demonstration units. The project was carried out by Østkraft, DTU, Danfoss, Vestfrost and Ea Energy Analysis.

3.3.7 Frequency Regulation Via V2G

At the University of Delaware in the eastern USA, research groups working on grid-integrated vehicles (GIVs) and vehicle-to-grid (V2G) power have undertaken a project where EVs provide frequency reserve services for the local TSO.

The central idea behind the project is that with the electrification of the vehicle fleet, an extremely large storage capacity in EVs will exist, a capacity that can be utilized to provide various ancillary services. This capacity exists owing to the fact that most EVs drive considerably less kilometres per day than the battery capacity allows for, and because EVs are parked roughly 22–23 h per day. This second point is also relevant with respect to the ability to provide ancillary services, as it means that, if plugged in, an EV is available to provide these services 22–23 h per day. In terms of effect, an EV is capable of producing over 100 kW. However, distribution grid and charging restrictions restrict this number. In the USA, the effect that can be delivered back to the grid is generally between 10 and 20 kW. In Denmark this figure is currently lower, at around 3.6 kW (single-phase 16 A connection); however, it is expected that three-phase 16 A connections capable of delivering an effect of 11 kW will be available to EV users in the near future.

The Delaware group highlighted three main components for their GIVs project which they developed (Kempton, 2010):

- A vehicle smart link that is tasked with controlling the charging, reporting to the server (capacity and current state) and logging/predicting future trips and times. This small unit was installed under the dashboard of the EVs.
- EV supply equipment which encompasses grid location, internet portal and power connections.
- An aggregation server to coordinate the real time operation of the EVs.

TSOs require a predictable and secure power resource to provide frequency reserve services; as such, the third of the above highlighted aspects, the aggregator, is a vital aspect of the project. While individual EV behaviour is not always very predictable, the aggregation of a large number of EVs is quite predicable, and thus provides a single, large, stable and reliable power source (Kempton, 2010). With this power resource at its disposal the aggregator can bid in on the frequency reserves market. The TSO does not have access to information regarding all the individual EVs, just the aggregate as a whole.

Delaware is part of the regional transmission organization (RTO), which also includes Illinois, Indiana, Kentucky, Maryland, Michigan, New Jersey, North Carolina, Ohio, Pennsylvania, Tennessee, Virginia, West Virginia and the District of Columbia. In this region, the frequency regulation is controlled by the regional transmission organisation via an automatic generation control (AGC) signal. Actors (typically generators) in the region submit hourly capacity bids with a minimum size of 1 MW; via the AGC signal, accepted bids are then activated by the RTO as needed. An accepted bid requires the actor to provide up- or down-regulation within 5 min, and an actor can be called upon to do so numerous times within an hour. While there is a small separate payment for the amount of electricity provided, the contract consists primarily of a capacity payment, with the actor receiving the same capacity payment regardless of how many times it is activated.

As part of the Delaware project, a number of legislative hurdles had to be addressed with respect to standards, incorporating the terms 'aggregator' and 'grid-integrated electric vehicles' into state laws, as well as issues relating to net metering, interconnection, and so on. In addition, the local distribution company has to approve each EV so they know how many vehicles will potentially be on each transformer. However, once all of this was in place, the aggregator could bid in on the frequency reserve market.

When the TSO informs the aggregator that regulation is required, the aggregator utilizes an algorithm that calculates how the load should be distributed and sends a 'request' to the various EVs. The algorithm is rerun every 4 s and, depending on the individual EV's reply, the dispatch is adjusted accordingly. For example, if an EV is not able to actually provide what it earlier indicated to the aggregator it could, the algorithm is rerun and a new dispatch is sent.

Project Results

So far the project has revealed that the power response has been very close to the command signal. As silicon is extremely quick to react, it has a higher fidelity than any rotating equipment and is thus extremely well suited to frequency balancing. Owing to the fact that the individual batteries are providing very small amounts of up- and down-regulation, the individual car batteries states of charge can be kept within an interval that does not cause excessive wear on the batteries. Financially speaking, the aggregator receives payment by the TSO and then passes on payments to the individual car owners.

3.4   Alternative Market Models for EV Integration

Today's marketplaces are built on large power production units, and the Nordic market in particular is built on large hydro power production units. This circumstance has led to market characteristics that fit this type of production. If, for instance, demand response had been the basis of the Nordic market place, then the market design probably would have looked much different from today. This section describes fundamentally different market approaches compared with the Nordic set-up (market splitting).

3.4.1   Locational Prices (Nodal Pricing in the Transmission Grid)

Locational marginal pricing (LMP), often referred to as nodal pricing, involves calculating market clearing prices for a number of physical locations (referred to as nodes) within a transmission grid. To calculate the nodal price, the generation, load and transmission characteristics for each node are required. Based on this, the nodal price represents the value of electricity in that node, including losses associated with distribution and congestion. This is quite different from the Nordic set-up, where electricity prices are calculated for an entire bidding area and congestions are handled through the division into price areas (market splitting). In both cases the transmission capacity is a given fact of life, which is taken into account in the energy pricing. While both are market based, nodal pricing factors are the costs of delivering electricity to a more specific area and more accurately reflect the cost of electricity at a specific point. However, this accuracy is obtained at the cost of a correspondingly higher complexity.

The idea behind nodal pricing is that the overall system balance is handled in the same iteration as grid congestions. Therefore, according to Stoft (2002), nodal pricing has two main benefits: (1) it minimizes the cost of production and (2) it provides the end-user with the true cost of their consumption. Seen from an EV standpoint, the main benefit may lie in the potential to be utilized in the distribution grids without causing congestions and at the same time help to obtain the overall system balance.

By providing the actual cost of electricity at each location, nodal pricing allows for a more efficient total dispatch and it sends price signals that better represent the cost in each node. As such, end-users are exposed to the true price of their electricity and, thus, can react accordingly.

Nodal pricing can also be designed to incorporate transmission losses into a local price, something that is not currently done in the Nordic system. As a result, if, for example, the electricity produced from a generator in northern Norway is just slightly cheaper than that produced in southern Denmark, this electricity could in theory be the marginal electricity that is sent to a consumer in southern Denmark. This is inefficient for the market as a whole, and savings could be realized if transmission costs were also reflected in local prices.

Locational prices are already utilized in a number of markets; for example, in New England. Although New England has 900 different pricing nodes, it only has eight different price areas, but the calculated nodal prices are used as shadow prices. Using only eight different price areas means less price differences between consumers. On the other hand, the precise prises and the fluctuation will not be fully utilized. If the consumers are very active, more than eight areas have to be considered.

3.4.2   Complex Bidding

In the Nordic electricity markets only a few and relatively simple bid types are used. For example, on the Nord Pool Spot market there are three types of bids available: an hourly bid, a block bid and a flexible hourly bid. Meanwhile, in the regulating power market, only a single bid type exists (X MW at Y DKK/MWh). This can be a challenge for a demand response such as the charging of EVs, since a change in charging (e.g. interrupting charging) will change the electricity demand at a later time (same day or next day).

Generators in the Nordic market determine their marginal costs of production individually and, based on these calculations, submit their bids. Nord Pool then aggregates these bids, which are benchmarked against the corresponding aggregated demand, and publicises the resulting price for every individual hour.

In other regions, more complex bidding forms are utilized, where generators must submit bids consisting of numerous parameters. Complex bidding is characterized by two factors that differ from the present Nordic market. First, the bid types are more technical and can be more detailed. Second, unit commitment is handled by the market, unlike in the Nordic market where the actors send in bid lists only with information about price and size, and where the unit commitment to some extent is calculated by the actor before sending bids to the market place. In New England, for example, generators include (Coutu, 2010):

- the economic minimum and economic maximum (dollars);
- up to 10 offer blocks (megawatts and dollars);
- minimum run and down times (fractions of hours);
  - when an asset is committed it must run for at least this amount of time before being shut down;
  - when an asset is de-committed it must be down for at least this amount of time;
- the no-load cost (dollars);
  - fixed cost incurred every hour the resource is running;
- the amount of energy that must be taken (megawatt-hours);

- the maximum daily energy available (megawatt-hours);
- ramp rates (megawatts per minute);
- notification time – time to start (hot, cold, intermediate);
- start-up costs (hot, cold, intermediate);
  - costs incurred per start-up of the resource.

In the day-ahead market in New England, based on the above inputs, as well as purchase bids and reserve requirements, the TSO carries out a system-wide unit commitment calculation at 12:00 the day before operation. Referred to as the resources, scheduling and commitment (RSC), this unit commitment calculation determines which resources should be committed to run for which hours, at a load somewhere between the resources minimum and maximum every hour, but the actual volume of the load is not determined at this time. During the next step, the scheduling, pricing and dispatch, this commitment schedule is run through an economic dispatch algorithm that respects the line and interface limits and determines the loading levels, demand clearing costs, final costs and locational marginal prices. This new commitment plan then undergoes a simultaneous feasibility test, and the constraints created are fed into a new RSC. When this cycle has been successfully completed, the day-ahead market is cleared and the day-ahead commitment and dispatch are published at 16:00. The result is financially binding schedules for demand and generation that are based on a least-cost security-constrained unit commitment, day-ahead hourly locational marginal pricing and binding constraints (Turner, 2010).

After the day-ahead market has been cleared and the results published at 16:00, there is a 2 h re-offer period that generally closes at 18:00. During this time, market participants can submit revised supply offers and revisions to demand bids for any dispatchable asset-related demand resources.

The vast majority of electricity market activity in New England is done so via the day-ahead market, where the bids and offers result in binding financial commitments. Deviations from these energy positions are dealt with in the real-time market, which is the balancing market. Offers in the day-ahead market, and revisions to these offers during the re-offer period, are carried over to the real-time market where they are combined with the actual load, actual external transactions and updated constraints. In the real-time market the dispatch is rerun and new locational marginal pricings are published every 5 min. After the closure of the operating day, finalized hourly real-time locational marginal pricings are also published. Real-time settlements are based on deviations between the day-ahead schedule and actual operations. See Figure 3.4.

Complex Bidding Summary

One of the major differences between the New England and the Nordic markets is that in the Nordic market, in terms of their bidding, actors must independently make an informed guess (and act strategically, for example, in the case of hydro power), whereas in New England all the relevant parameters are delivered to the TSO, who then optimizes the dispatch and unit commitment. Owing to the fact that the TSO has access to more information in the New England market, and less is left to educating guessing, it is expected that the co-optimization

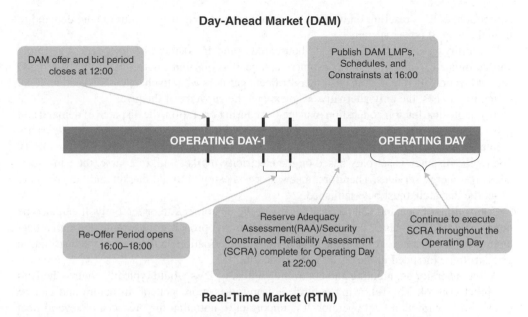

Figure 3.4 Aspects of the New England day-ahead market (DAM), and real-time market (RTM) (Turner, 2010).

that is undertaken will lead to a more efficient dispatch. The ability to include additional parameters in the bidding process, as well as the direct coupling of the day-ahead market with the real-time market, is well suited to more complex offers; for example, 'energy storage' or 'demand response', two aspects that would be particularly relevant for EVs.

3.5 Management of Congestions in the Distribution Grid

In a future with a large share of wind power, and many heat pumps and EVs, the capacity of the distribution grid may be challenged. The increased electricity consumption from heat pumps and EVs can in and of itself create problems with insufficient capacity in peak hours in some distribution grids. In addition, the combination of a large electricity consumption and price-controlled demand may in some cases create problems. This is the case if the demand's reaction on wholesale market prices implies larger movement of demand than the distribution grid can accommodate.

The following discussion only considers slower charging of EVs where the EVs are connected to the grid and it is possible to optimize the charging, for example, in accordance with a price signal. Other ways to charge an EV, such as fast charging and/or battery swapping, are not considered.

Traditionally, demand takes place when electricity is needed and the challenges in the distribution grid caused by demand spikes are solved by expanding the grid to fit the size and patterns of demand. As an alternative – if the demand is flexible – demand peaks can be reduced (i.e. moved to another hour of the day to avoid short-term overloading of grid components). EVs reacting on wholesale market prices is one way of introducing demand to fit in the overall power system.

Flexibility in demand from private households (and EVs) must be included in a simple and automated manner so that consumers can easily contribute with flexibility. If demand cannot be controlled (e.g. if no customers or fleet operators want to change plans) and no local generation exists, the only alternative is to expand the capacity of the grid.

To enable distribution congestion management during operation, the dispatch of demand and generation must not only be based on the power system balance (the conventional electricity wholesale market), but also on the operational state of the grid (information from the DSO). Today, the dispatch is set only based on the electricity market and, of course, the end-users' needs for energy services. Therefore, a new process, using both the market and the grid state to set the dispatch, must be established.

Controlling of demand can be done locally at the customer site or remotely. It can also be done using price signals or using a more direct control approach; for example, through a fleet operator. For this control of demand to be effective, information on the state of the distribution grid has to be included in the basis for EV reaction.

A fleet operator aggregating and controlling a fleet of EVs would typically control the EVs by direct control. The following control description is more general in nature and can be used with or without a fleet operator. It is important to note that the presence of several fleet operators will be necessary in several of the methods described; this is because, if only one fleet operator reacts, statistical methods to predict the consumption and effect cannot be used.

As a minimum, the following conditions have to be met:

- the demand can be controlled;
- information representing the state of the distribution grid has to be included in the basis for reaction of the EV;
- introducing price-based methods requires rational behavior from the end-user (through the retailer, etc.).

3.5.1   Section Overview

The following sections describe several methods to handle and avoid overload of the distribution grid, with all taking a point of departure in the requirement for voluntary customer participation. Some of the approaches are compensation based ('carrot') and some are penalty based ('stick'). Other methods based on more or less involuntary participation (e.g. grid-code based or DSO terms of delivery) are, however, not considered here.

First, the obstacles that have to be overcome to handle congestions in the distribution grid from the DSO point of view are described. Second, general thoughts regarding how to handle grid congestions are described. Next, principles, advantages and disadvantages to handling congestions in the distribution grid are described for the following types of solutions:

1. payment for the right to use capacity
2. time-of-use tariffs
3. direct control ('cycling')
4. a bid system

5. dynamic distribution grid tariffs
   i. spot price
   ii. spot and regulating power price
   iii. actual power consumption.

Last, a mixed approach suited for a VPP (equal to the sum of several EVs) which lives up to minimum requirements so that the approach fits the current system without the need for changes is described. It considers EVs both in the spot and in the regulating power market, and is relatively socio-economic and relatively simple.

3.5.2   The Role of the DSO

In the transmission grid, congestions in the spot marked are handled through a price signal. This implies that, if congestions exist, the spot price increases in the area behind the congestion (and the cheaper bid from the area in front of the congestion cannot be used). In the regulating market, congestions are handled by jumping over a bid in the merit order list, implying using a more expensive bid. In principle, the TSOs 'remove' congestions by grid reinforcement in the transmission grid if a socio-economic business case can be shown. This means that the TSOs invest if the price for removing the congestion (via grid expansions or technical solutions) is less than the cost 'punishment' in the spot market and the regulating power market for not using the cheapest bids.

The question then becomes, can the same be done in the distribution grid? The answer in principle is yes; however, in practice, several obstacles have to be overcome.

Mapping Congestions in the Distribution Grid

In order to identify the congestion it is necessary that the distribution company has full insight in the distribution grids. This will require investigation and implies investments in measurement equipment. It is also necessary that the distribution company identifies the long-term marginal costs of reinforcement of the grid in order to be able to compare these costs with the cost involved to make end-users shift demand in a given period. The comparison of these two figures can give a signal of when to reinforce and when to reduce/move demand dependent on which is most cost effective for the DSO.

Depending on how demand-side response is implemented and to what degree the response is voluntary or not, the cost for the distribution company may include an additional overhead to provide an incentive for the end-user to shift demand.

Geographically Differentiating

If conditions in the distribution grid have to be taken into account, finding a way to include a signal representing the distribution grid will be necessary. This signal is by its nature both time and geographically dependent. However, legislation often restricts distribution companies from making geographically differentiated prices. As such, legislation changes may first have to be undertaken before several of the described methods implying geographical prices can be utilized.

In addition, it is also a challenge to explain to end-users that two houses close to each other have to pay different prices for charging an EV, strictly due to past grid extension decisions.

Development of new DSO Tools for Price-Dependent Demand

Generally speaking, electricity demand in most countries is not flexible. Models for price dependency in demand, therefore, remain to be developed and included in all aspects of the balancing of the energy system. If the DSO is to play an active demand-controlling role then the DSO has to develop models for the costumer's price-flexible behaviour. Alternatively, the demand-side response can be handled by the aggregators/retailer/fleet operators, who know their customers the best, and the DSO may only provide information about the capacity available for flexible consumption in the distribution grid. In both cases the contractual and legal framework has to be in place to specify rights, obligations and pricing mechanisms for all parties involved.

The models for price dependency in demand take their basis from historical data. Using measurements of historical demand might be more difficult at lower grid levels (and there are also fewer customers, and thereby a smaller dataset).

Pricing by the DSO

An important prerequisite for a business case is the existing economic regulation of the DSO. The regulation should allow for both (i) full cost-recovery of the grid investments, including return on invested capital, and (ii) flexible tariffs and compensation schemes allowing the DSO to provide its services on the basis of the long-run marginal costs. The latter will probably lead to increased tariff differentiation, which, as was highlighted above, may be somewhat legally and/or politically problematic.

3.5.3   Overall Approach: The Order of System Balance and Grid Congestions

Handling of grid congestions can be more or less integrated with the handling of the system balance in the overall energy system. In addition, handling of congestions in the distribution grid can more or less be market based. This section looks at different approaches based on their order and the degree of integration in the handling of grid connection and system balance. The three categories are:

1. Integrated process.
2. Stepwise process: first system balance, then grid congestion.
3. Stepwise process: first grid congestion, then system balance.

In addition, one could also consider iterative processes or a mix of the above processes.

Integrated Process

In an integrated process the dispatch of demand and generation according to system balance and grid constraints are set at the same time and/or in the same procedure. A principal example of an integrated process (at transmission level) is the current Nordic spot market where the system balance and congestion of inter-area connections are handled in a market splitting process. Such an integrated process is one (and maybe the only) way to get an optimal or close to optimal socio-economic operation in all situations.

The disadvantage of an integrated process can be that it is complex to understand and perform. In addition, if distribution grid congestion should be managed to be integrated with the global system balance, it will require the handling of a huge amount of grid nodes in one optimization process.

The largest challenge might be that an integrated process requires modification of the current electricity markets (or even introduction of a completely new approach). This will not be an easy task, as the current market is a common Nordic market moving towards a common European market, which means that lots of coordination work has to be done. It would take a number of years to obtain a common European solution.

Nonetheless, examples of possible implementations include:

- Nodal pricing (LMP).
- Hierarchical market (aggregation) aligned with the grid topology. The DSO can interact with the local market levels to avoid congestions; for example, through a bid-based system (local adder bids to avoid grid constraints).
- Control-by-price based system with locational adder prices to avoid grid constraints.
- Market for grid capacity (seller: DSO; buyer: end-users; the selling price curve will be steep around the capacity limit) operating in parallel with the spot market.

Stepwise Process: First System Balance, Then Grid Congestion

In this stepwise process the system balance is handled in a separate step before handling of grid congestions. It is expected that (the main part of) the dispatch has to be handled in the first system balance step. This is reasonable owing to the energy volumes involved in the system balancing and an assumption that grid congestions will not be the normal operating situation.

This dictates that the grid congestion management has to be an adjustment to the dispatch in the first step. As an example, the system balancing is handled in the current spot market, which may lead to a dispatch of demand and generation where the grid is overloaded at critical locations and times. This overload can be handled in a second step; for example, a market where the DSO can activate specific bids provided by the users, and thereby relieve the overloaded parts of the grid. Also non-market-based methods (direct control) can be used; for example, direct control of demand/generation based on contractual agreements between the DSO and retailer.

The advantage of a stepwise process – with first system balance, then grid congestion – is that it is relatively easy to understand. That the solution can be made as an extension of the current market system is also an advantage, as the current wholesale markets do not have to be modified.

The disadvantage of a stepwise process is that an optimal solution cannot be obtained, especially in cases where grid constraints are so large that the normal market is influenced. In principle, the system balance has to be run again and then the method in the next section can be considered.

Stepwise Process: First Grid Congestion, Then System Balance

In this stepwise process the grid congestion is handled in a separate step before handling of the system balance. The dispatch is expected, as today, to be set in the step that handles the system balance. The very simple (and probably not very efficient) example of such a process is that the DSO in the first step announces capacity limitations to all end-users (through fleet operators or retailers), and that the consumption in the second step trades in the current market within these capacity limitations.

A more sophisticated (but still simplified) process in line with this approach could be that the DSO first sets a grid tariff for each time slot and each grid node for the coming day of operation – like dynamic tariffs. Second, the end-users bid in the current spot market taking into account the dynamic grid tariff. If congestion is expected to take place, the DSO will set an expensive tariff for the specific time slot and location. This will lead to a tendency that demand does not bid in the time slots involved, and thereby grid overload is avoided. If local generation exists, then the tariff will also send a signal to produce when needed. One must note that this concept includes a time variable and geographically dependent tariff.

The advantage of a stepwise process – with first grid congestion, then system balance – is that it is relatively easy to understand. That the solution can be made as an extension of the current market system is also a quite large advantage, as the current markets do not have to be modified.

The disadvantage of this type of a stepwise process is again that an optimal solution cannot be obtained, especially in cases where grid constraints are so large that the normal market is influenced. Another disadvantage of taking the grid congestions first is that it can be difficult to set incentives at a sufficiently accurate level to avoid congestion.

The following sections look at different types of approaches in more detail.

3.5.4   Payment for the Right to Use Capacity

Today, consumers are entitled to draw a high effect from the distribution grid. This might not be suitable if grid congestions exist. A special payment could be introduced for users in need of more than a certain maximum. An example of this exists in Italy, where the standard limit is 3 kW per household. This system is simple and also easy for the consumers to understand. It requires a form of local control of demand; for example, making the charging of the EV dependent on the actual consumption by the rest of the household.

The capacity payment could be the same year around or could be adjusted by season to reflect capacity scarcity.

It is important that the payment is set so that capacities are not exceeded at any time. This means that local optimization within every household is made. This is a nonflexible solution that will probably not lead to a cost-efficient solution, as reduction at the individual level does not always coincide with system peaks. If the payment was configured for long periods (e.g. months), then capacity would also be expensive in periods with no capacity problems (e.g. nights).

A more optimal solution could be to split consumption into nonflexible and flexible consumption, and only announce capacity limitations for the flexible consumption. This requires a fleet operator with a large portfolio of EVs so that they, within the limitations, can, for example, load all the EVs in an optimal manner either by sending price signals to the end-users or an on/off signal. In this approach, the grid congestions are handled separately from the system balance and the wholesale market. However, there is still a connection via the payment mechanism.

3.5.5 Variable Tariffs (Time of Use)

Time-of-use-type grid tariffs (fixed day/night/weekend tariffs) can be used in the management of congestions in the distribution grid in peak hours. Time-of-use tariffs could shift certain habitual consumption (e.g. the start of a washing machine) from costly to less-expensive hours. Time-of-use tariffs can be used by retailers, but in this context the distribution tariff is brought into focus.

With time-of-use tariffs the end-user will be able to react; for example, by using a simple timer device. However, time-of-use tariffs do not support any variable response and, therefore, they cannot be used to avoid price-introduced peaks (e.g. activating down-regulating power – increased demand – in the night). The time-of-use tariff will provide an incentive in the right direction, but it will not be enough to solve grid congestion at all times. It might lead to constraints in the distribution grid, particularly in low-tariff periods. There is also a risk that, instead of smoothing out demand, the peak is simply moved to another hour.

3.5.6 Progressive Power Tariffs

A simple method could be to set up fixed tariffs depending on the actual power consumption, also known as power tariffs. The progressive power tariff is fixed and has a known relation to the actual power consumption. High power consumption is equal to a high tariff and low power consumption is equal to a low tariff. This creates incentives for customers to smooth out their demand throughout the day. Therefore, tariffs based on the actual consumption increase the amount of energy that can be transferred in the distribution grid, especially in the low-voltage grid.

Power tariffs do not directly support the overall balancing of the system. This can be done if the fixed tariffs are used in combination with other market models; for example, end-user participation in the markets for ancillary services (i.e. regulating power market or market for reserves).

The advantages of the method are that it is simple and relatively easy for the DSO to implement because all customers have the same tariffs and the tariffs are fixed and known beforehand.

The method has no direct link to the system balance; but, assuming that the electricity demand curve expresses the need for balancing and the (over) load in the grid, there is an indirect link. When the shape of the actual consumption expresses the need for balancing, it is a well-suited method. This is often the case today where spot prices are strongly correlated to demand (i.e. prices are normally high on weekdays between 16:00 and 19:00 when demand is high). However, in a future with much more wind power this method will be more inaccurate owing to the need to integrate this increased wind power in the overall system (i.e. demand may still be high between 16:00 and 19:00, but when winds are strong during these hours the price may not necessarily be high).

3.5.7 Direct Control: Regulatory Management

Another way to manage congestions in the distribution grid is to allow direct control from the grid company. This can be done by reducing the electricity consumption for selected units (e.g. EVs and heat pumps) by use of remote control. The control can be used more systematically in a cycle to avoid continuous overload.

One example comes from the USA, where cycling is often used for air conditioning. When load must be reduced, air conditioners are only permitted with an on/off cycle of, for example, 50% on. The fraction of on-periods and the payment can be adjusted to balance the need and willingness for customers to participate. The US system is often applied to solve large-area power balancing. The cycling of consumption is only activated when needed, so this type of control can be said to be accurate. However, the system is often only activated in relation to one or a few standardized technologies (e.g. air conditioning). Other types of demand – that also could hold potential for control – are not considered.

Another example, often referred to as 'brown outs', can be used at system level to comply with lack of generation capacity in certain peak load situations.

The same concept could be applied for management of congestions in the distribution grid. This would require a signal – geographically dependent to the level of interest – based on market prices or technical measurements; for example, the (over)load of the grid.

The market aspect of this method is related to a possible introduction of a compensation mechanism. Ideally, the compensation should be divided to a level that takes into account the risk of congestion and should also reflect the potential reduction in effect and/or time. The compensation can be calculated as the difference between the wholesale market prices in the hours from where the demand is moved from and to, plus a loss from 'pain and suffering'. The compensation will reflect the alternative cost for expanding the grid; that is, if the payment is too high, the grid company will reinforce the grid instead. One should note that deciding the compensation is not straightforward, as it demands full insight in the grid and grid costs.

Under this approach, the grid congestions are handled separately from the system balance and the wholesale market. There will, however, still be a connection expressed through the compensation mechanism.

3.5.8 Bid System

In the Nordic transmission network, a bid system is used to manage congestions in the spot market. Based on bids for generation and demand, the prices are calculated so the transmission capacity is used to its maximum. This is supplemented with another type of bid system in the regulating power market where generation and demand bid into a merit order list. Starting with the cheapest bids, the TSOs activate bids when needed. Congestions are handled by leaving a bid out from the merit order if it cannot be used because of congestions in the transmission grid.

In theory, the same concept could be used in relation to distribution grids. For example, individual customers or fleet operators could send bids to the grid company describing the amount of demand they can reduce and the requested price in the relevant geographical areas. When needed, the grid company could activate the cheapest bids taking into account the geography. A bid system could also be used to handle automatic reserves.

In such a concept, many bids concerning demand reductions (or generation) are required, and the system would result in individual prices in all parts of the distribution grid. In contrast to the transmission grid, this market type for the distribution grid would have few participants. The risk, therefore, is that the competition may be very small, potentially resulting in gaming.

The challenge is to design a bid system that gives incentives for smart grid investments rather than grid reinforcements.

3.5.9   Dynamic Distribution Grid Tariffs

Dynamic tariffs reflecting the underlying marginal costs can give a price signal to end-users about the real costs in the total system from the congestions induced by the specific end-user. Such a dynamic tariff varies in time and location. A basic principle for dynamic tariffs could be to let the tariffs reflect the marginal cost by the increased demand (or generation) at a specific time and in the specific part of the grid. A dynamic tariff will give end-users an incentive to adjust their demand/generation depending on the costs of their contribution to their grid load. As a result, end-users who get price signals based on marginal costs get a clear incentive to act efficiently – both from a socio-economic and an individual point of view.

If the socio-economic cost of grid reinforcement is lower than the socio-economic cost of end-users reacting flexibly, the grid shall be reinforced. It should be noted that this requires that the DSO can establish relatively straightforward business case calculations (see Section 3.5.2), in order to be able to incorporate the short-term flexibility in the long-term planning (the complexity of this should not be underestimated).

A working group under the Danish Energy Authorities has made a report investigating the possibilities for and impacts of dynamic tariffs (Energistyrelsen, 2010). The report states that the DSO's task is, as far as possible, to identify their long-term marginal costs and establish tariffs according to that. When the tariffs reflect the long-term marginal costs, the end-users' decisions will ensure efficiency in the costs for both customers and DSO. The DSO will not reinforce the grid unnecessarily and the end-users will not pay more than the shadow price for increased grid capacity. It should be noted, however, that it can be somewhat problematic that the long-term marginal costs per definition are average condition and that dynamic tariffs are short-term hourly figures. Developing a method for when to invest must include a connection between those two figures.

The set-up could consist of the following elements:

- It is accepted that the payment for distribution grids is dynamic and may vary over time and location.
- Normal grid tariff is paid in locations without congestions.
- When capacity problems are expected, the grid company broadcasts an additional tariff in each hour for each location. The size of the tariff should be just large enough to solve the capacity problems (before or after handling of the system balance).

• When the profit from this congestion management exceeds the periodic costs of grid extensions, the grid extensions will be carried out.

The price level is dependent on a balance between the size of the capacity problems and the costumers' willingness to reduce their electricity consumption. The profit from the transitory high grid tariffs could be used to reduce the tariff in the affected areas in periods with no constraints, so that the average tariff is the same in all areas. This might not be straightforward, and methods for this have to be developed. The dynamic tariffs representing grid congestions could be on top of the ordinary grid tariffs – if no congestions, then the dynamic tariff is zero. This could, for example, be calculated through an allocation between ordinary demand and flexible demand.

Dynamic Distribution Grid Tariffs: Spot Price

The following method takes as its point of departure a distribution grid tariff that interacts with the spot market.

Generally speaking, when capacity problems are expected, the grid company would broadcast an additional tariff the day before, for each hour and for each location. The size of the tariff should be just large enough to solve the capacity problems. The method implies that the spot price and the grid tariff are set the day before and can, therefore, interact, and the costumer can react on the sum of these two price signals. The signal received by the consumer consists of two parts (excluding taxes, etc.):

$$\text{Total Price} = \text{Spot Price} + \text{Distribution Grid Tariff}$$

The method is relatively simple, as it does not imply that customers must submit bids or plans for their expected electricity consumption, and real-time measurements of consumption are not required. The method is quite similar to the way congestions are handled in the transmission grid, but is somewhat simplified. The expected consumption has to be estimated by the grid company, and a model based on historical data could be used to estimate this expected consumption. The challenge is to create a reliable set of data for a small number of consumers. The lower the level (and the fewer customers) in the geographical area the more difficult it might be to obtain a robust dataset.

Whether or not to handle the congestions in the distribution grid before or after the spot, or in a mixed approach, will depend on whether the volumes are sufficiently small such that imbalances can be ignored. If the imbalances can be ignored, the grid congestion can, for example, be handled before spot, and the tariff can be published after spot prices are announced. If imbalances cannot be ignored then the tariffs must be somewhat predicted before 12:00 the day before operation, and indirectly included in the bids to the spot market.

Dynamic Distribution Grid Tariffs: Spot and Regulating Power Prices

If the EV (or demand in general) reacts on spot prices set the day before, it will be sufficient as described above to react on grid congestions once a day. But if EVs also are to react on continuously updated regulating power prices, it can cause grid congestions if the grid constraints are only considered once a day. For example, an unexpected wind front would cause very low (e.g. negative) prices in the regulation power market. This could motivate price-signal-based consumption units to increase their demand (e.g. EVs start charging). If grid constraints are not handled at the same point of time, a risk of overloading distribution grids exists. This implies that the signal received by the consumer must consist of three parts (excluding taxes, etc.):

$$\text{Total Price} = \text{Spot Price} + \text{Regulating Power Price} + \text{Distribution Grid Tariff}$$

In the case of EVs reacting on both spot and regulating power price the distribution grid tariff will have to be updated continuously with updates based on the regulating power price. In comparison with the method above using solely spot prices, this is more complicated as consumption has to receive updated price signals several times a day. However, the method still does not imply that the customers have to submit bids or plans for their expected electricity consumption.

As in the case of solely using the spot price, real-time measurements of consumption are not required. The expected consumption has to be estimated by the grid company. Once again, a model based on historical data can be utilized and the same challenge exists with respect to creating a credible dataset.

3.5.10 Comparison

As was outlined above, several solutions for grid congestion management can be considered. The characteristics of the various methods are shown in Table 3.2.

Note that all the methods shown are based on voluntary customer participation. Other methods based on more or less involuntary participation (e.g. grid-code based and/or DSO terms of delivery) are not considered here.

Some factors that are important in the selection of which method to use include:

- Whether the solution can be integrated into the current energy market concepts with limited modifications (or at least an evolutionary development of existing market concepts into a new approach should be possible).
- The solution must be able to handle different types of actors, commercial and noncommercial, as well as several commercial actors (traders, VPPs, etc.) responsible for customers in the same grid location.
- To the greatest extent possible, the solution should include socio-economic aspects in the dispatch and decisions regarding grid reinforcements.
- The solution should be easy to understand for end-users.

Management of congestion in the distribution grid that takes into account end-users' reactions to price signals is a new and untested area. All methods require more knowledge of the distribution grid than is available today. In addition, new methods to handle congestions in the distribution grid and to predict the reaction of the consumers have to be developed. In particular, mapping of congestions in the distribution grid, geographical differentiation of end-users, pricing through compensation mechanisms and the development of new tools for including

Table 3.2 Characteristics of various methods.

Aspect	Solve traditional peak load and/or price induced peak?	Accuracy: Does reduction of demand only occur when needed, and with a reasonable amount?	Simple or complex	Comments
Payment for the right to use capacity	Both	Not accurate	Simple	Would affect demand even at times without capacity constraints
Direct control ('cycling')	Both	Accurate	Simple	Typically only for a limited number of technologies
Time-of-use tariffs	Traditional peak	Not accurate	Simple	Most useful for behavioural change
A bid system	Both	Accurate	Complex	Can be too complicated to manage by individual end-users. A fleet operator may be needed
Dynamic tariffs (in relation to spot price)	Traditional peak	Accurate	Simple	Demand is relatively easy to predict, even if influenced by spot prices
Dynamic grid tariffs (in relation to spot and regulating power price)	Both	Accurate	Complex	The combination of high price incentives in the regulating power market and that, for example, very low prices can occur at any time is a challenge

price-dependent demand are obstacles to be overcome before distribution grid congestions can be handled in a new manner.

3.5.11 Operation of a VPP for EVs

Another method for dealing with grid congestions is via a VPP. A scenario with power system balancing (spot market and regulating power market) and grid congestion management (dynamic tariffs) could look as follows:

1. Grid congestions in the coming 24 h are forecasted and tariffs are set in specific geographical areas and hours (dynamic tariffs).

2. VPP bids in spot market taking (i) forecasted spot market prices and (ii) grid tariffs for each hour and each location into account (for each hour and location: total price = forecasted spot market price + grid tariff).
3. Spot prices are set by the spot market and the VPP overall fleet charging schedule will be established based on activated bids.
4. The VPP establishes individual charging schedules and distribute them to EVs.
5. The VPP and other retailers offer up- and down-regulating power in different grid locations previously appointed by the DSO.
6. During operation, grid constraints may occur due to deviations of demand and generation from the plan (or 'wrongly' set tariffs by the DSO).
7. The DSO activates location-specific regulating power offers from step 4 if congested lines occur ('counter activation' may be needed to avoid the influence of the system balance, and a method for coordination with the TSO and the conventional regulating power market for balancing must be developed).

This method is a combination of dynamic tariffs for spot market timeslots, with a first system balance/then grid congestion for regulating market timeslot. Steps 5–7 can be replaced with a control-by-price approach.

References

Coutu, R. (2010) Unit commitment and dispatch. Presentation, ISO New England.

Demonstration Project (2011) Electricity demand as frequency controlled reserves, 2011, http://www.ea-energi analyse.dk/projects-english/927_electricity_demand_as_frequency_controlled_reserve.html (accessed February 2013).

E-bridge (2011) Analysis & review of requirements for automatic reserves in the Nordic synchronous system. Intermediate Technical Report. E-bridge, 21 September.

Energinet.dk (2010) Udvikling af rammer for regulerkraft – Indpasning af mindre forbrugsenheder og andre mindre enheder i regulerkraftmarkedet (Development of a framework for regulating power – incorporation of minor consumption units and other smaller units in the regulating power market), May.

Energistyrelsen (2010) Redegørelse om mulighederne for og virkningerne af dynamiske tariffer for elektricitet, June 2010.

Kempton, W. (2010) The grid-integrated electric vehicle. Presentation at Electric Vehicle Integration into Modern Power Networks course, Technical University of Denmark, Department of Electrical Engineering, Lyngby, Denmark, September.

Nordel (2008) Harmonisation of balance regulation in the nordic countries, December.

Stoft, S. (2002) *Power System Economics: Designing Markets for Electricity*. John Wiley & Sons, Inc., New York, NY.

Turner, D. (2010) Energy markers: overview of the day-ahead marker (DAM). Presentation, ISO New England.

4

Investments and Operation in an Integrated Power and Transport System

Nina Juul[1] and Trine Krogh Boomsma[2]
[1] *DTU Management Engineering, Technical University of Denmark, Roskilde, Denmark*
[2] *Department of Mathematical Sciences, University of Copenhagen, Copenhagen, Denmark*

4.1 Introduction

Electric-drive vehicles (EDVs) are often considered another load to the power system. Furthermore, investments in EDVs are expensive. Analysis of optimal investments in the power and road transport system separately results in EDVs being too expensive and only some investments in sustainable energy on the power system side. However, analysis of optimal investments and operation of an integrated power and transport system results in EDVs being the optimal choice for investments, which in turn allows for a larger share of sustainable energy in the power system.

With an increasing amount of renewable energy sources, the power system faces the challenges of including sufficient flexibility in the remainder of the system. The EDVs can provide a large amount of flexibility, as well as enable peak load shaving and regulating power. With focus on only the transport system, it is not beneficial to invest in EDVs. However, with inclusion of the benefits of the power system, the EDVs become beneficial from a socio-economic point of view.

A number of aspects are related to integration of the power and road transport systems. Research has been done within various fields, such as potential benefits for the power system and for the customers, infrastructure, transition paths, and quantification of the impact and benefits. Kempton and Tomic (2005a) gave an introduction to the concept of vehicle-to-grid

Grid Integration of Electric Vehicles in Open Electricity Markets, First Edition. Edited by Qiuwei Wu.
© 2013 John Wiley & Sons, Ltd. Published 2013 by John Wiley & Sons, Ltd.

(V2G) and potential benefits of V2G. This article was followed by an article including business models as well as a discussion on dispatch of vehicles (Kempton and Tomic, 2005b).

Many articles have followed, and interest has increased within this field. Focus on potential benefits of particular services has been taken in terms of peak load shaving in Japan (Kempton and Kubo, 2000), and regulation and ancillary services (Tomic and Kempton, 2007). Brooks (2002) has looked into integration of battery electric vehicles (BEVs) with particular focus on the benefits of the vehicles providing ancillary services. Cost comparisons of providing different kinds of services have been made by Moura (2006), comparing the different kinds of EDVs with the technologies providing the services today. In general, these papers find that it is beneficial to introduce EDVs, although the economics of using these for only peak load shaving (in Japan) is not very profitable without a change in the electricity price schedules (Kempton and Kubo, 2000).

So far, the impact on power production from an integration of the power and road transport has been quantified by few. McCarthy *et al.* (2008) have developed a simplified dispatch model for California's energy market to investigate the impacts of EDVs as part of the energy system. However, this model does not take the fluctuation in power production and, thus, the need for flexibility into account. Kiviluoma and Meibom (2010) compared power system investments and CO_2 emissions in scenarios with different amounts of wind, heat pumps and plug-in hybrid electric vehicles (PHEVs) in the Finnish power system. However, this paper does not include investments in the vehicle fleet. Lund and Kempton (2008) have made a rule-based model for the integrated power and transport system, focusing on the value of V2G with different levels of wind penetration.

Juul and Meibom (2011a) contributed with the investment model for the integrated power and road transport system presented in more detail here. This model makes it possible to analyse the impact of interaction on future investments in far more detail and potentially provides more insight on the changes and benefits in the power system than in the papers mentioned above. The calculation of investments in different vehicle types has not been included in any of the above, but has been introduced by Ibáñez *et al.* (2008) in terms of illustrative cases, and Juul and Meibom (2011a) illustrated the use of the detailed transport system model. Juul and Meibom (2011b) showed how the introduction of electrical power in the transport sector and investment decisions regarding vehicle types and power system configuration have consequences for the entire power system; for example, in terms of optimal mix of production and storage units, fuel consumption, and CO_2 emissions.

This chapter builds on studies in the articles by Juul and Meibom (Juul and Meibom, 2011a,b) and the PhD thesis of Juul (2011).

4.2   The Road Transport System

Many road transport systems are characterized by almost only internal combustion engine (ICEs) vehicles among the passenger vehicles; for example, the road transport system in Denmark. The number of hybrid electric vehicles and BEVs with quite small batteries are increasing, but, as of today, electricity only counts for 0.01% of the passenger road transport in Denmark (Danish Statistics, 2010).

4.2.1 Expectations of Future Road Transport System and its Integration with the Power System

Moving towards large shares of renewable energy calls for a shift in road transport. Renewable energy in the transport sector can be bio-fuels, electricity or fuel cells. Bio-fuels are available in limited quantities and are not of interest for the power system. Thus, the focus of this chapter is EDVs, including their integration with the power system. The types of EDVs in focus are:

- **BEVs:** These vehicles drive on electricity only. No range extender in terms of an engine or fuel cell is included. Hence, BEVs have a limited driving range before charging the battery.
- **PHEVs:** These are vehicles with a battery for driving short distances, in cities or the like. Besides the battery, the vehicles have an engine; for example, using diesel as a propellant. The engine can be regarded as a range extender, ensuring that the PHEV can drive long distances.
- **FCEVs:** Fuel-cell plug-in hybrid electric vehicles (FCEVs) have a battery like PHEVs. Besides the battery, FCEVs have fuel cells using, for example, hydrogen as a propellant.

Furthermore, ICE vehicles are considered in order to analyse optimal investments. These are defined as vehicles with an engine that uses either diesel or gasoline as propellant, like most of the vehicles on the roads today.

4.3 The Energy Systems Analysis Model, Balmorel

The Balmorel model was initially developed to support energy systems analysis in the Baltic Sea region, with focus on both electricity and combined heat and power sectors (Ravn *et al.*, 2001). In the Balmorel model, experts, researchers or the like can analyse aspects with regard to economy, energy or environment, including future configuration of the energy sector within a region or in all the countries analysed.

Balmorel is a deterministic, partial equilibrium model assuming perfect competition (Ravn *et al.*, 2001). The model maximizes social surplus subject to (a) technical restrictions (e.g. capacity limits on generation and transmission, and relations between heat and power production in combined heat and power plants), (b) renewable energy potentials in geographical areas and (c) electricity and heat balance equations. The energy sector experiences price-inflexible demands; thus, maximizing social surplus corresponds to minimizing costs. Balmorel generates investments and dispatch that results in an economically optimal configuration and operation of the power system. Market prices for electricity can be derived from marginal system operation costs. Input data changes allow for sensitivity and scenario analyses.

In Balmorel, countries are divided into regions connected by transmission lines. Regions are then further subdivided into areas. Electricity and transport demand is balanced on a regional level, whereas district heating is balanced on an area level. Balmorel has a yearly optimization horizon, with an hourly time resolution. Long-term investments are typically run with either aggregated time steps or weeks. For some scenarios an hourly time resolution is important; therefore, an aggregation of weeks is used. In scenarios with a large share of renewable energy sources, the hour-by-hour variations are important. Otherwise, the required flexibility is underestimated. Furthermore, use of a yearly time horizon implicitly assumes that conditions do not vary year by year. This is a rather crucial assumption, as it will be computationally intractable to include every hour of, say, a 20-year horizon.

The investment decisions in the model are based on demand and technology costs, including annualized investment costs given the particular year. Investments in Balmorel can be both endogenous and exogenous. If investments are exogenous, only the operation of the power system is optimized under the given power system configuration. Furthermore, some capacity can be given beforehand, enabling the model to optimize remaining investments based on the given configuration. Data for the existing heat and power system is given on a detailed level, whereas investments are given on plant technology level in each region (for heat production in each area).

The Balmorel model is modelled in a way that allows for add-ons (Ravn, 2010). That is, inclusion of more specific models for an area that needs special investigation in relation to the rest of the energy system is possible. The add-on will become part of the model, and, thus, when chosen active, the energy system model and add-on can be considered to be one model. Add-ons have previously been made for analysis of, for example, hydrogen and detailed waste treatment (Karlsson and Meibom, 2008; Münster and Meibom, 2011). The EDV model presented in this chapter is an add-on for Balmorel.

4.4   The Modelling of Electric Drive Vehicles

Based on input data, the Balmorel model with inclusion of the road transport system minimizes total costs. The model has to meet constraints on transport demand and power flow balancing. The interaction of the transport system with Balmorel includes addition of net-electricity use for transportation to the electricity balance equation of the entire energy system (the electricity use less the power fed back to the electricity grid). The output of the model is an optimal configuration and operation of the integrated power and transport system.

The transport system model, including transport demand, vehicle technologies and V2G capabilities, is developed as an add-on to Balmorel. This section elaborates on some of the details of the transport add-on, described in Juul and Meibom (2011a). Extending the Balmorel model with the road transport sector enables analysis of:

- The economic and technical consequences for the power system of an introduction of the use of electricity in the road transport system, either directly in EDVs or indirectly by production of hydrogen or other transport fuels.
- The economic and technical consequences of an introduction of V2G technologies in the power system; that is, BEVs and PHEVs being able to feed power back into the grid.
- The competition between different vehicle technologies when both benefits for the power system and for the investment and fuel costs of the vehicles are taken into account.

The transport model includes demand for transport services, investment and operation costs, and electricity balancing in the transport system. In this first version, only road transport using vehicles for passenger transport is modelled. Inclusion of other types of road transport services (e.g. transport of goods) in the model is a matter of data availability and collection.

4.4.1 Assumptions

In modelling the EDVs, it is assumed that the infrastructure is in place. Moreover, the following assumptions have been made:

- Vehicles are aggregated into vehicle technologies; for example, a limited number of BEVs are used to represent all types of BEVs.
- For each vehicle technology, batteries are aggregated into one large battery for all vehicles parked. This has no influence on the outcome of the model. However, if we would like to optimize the use of each vehicle, the batteries need to be considered separately. This would, however, result in exhaustive calculation times.
- The transport pattern is considered with average hourly values on a weekly basis. The transport patterns are assumed known, which makes it possible to extract average values (see Section 4.5.2). Transport patterns are averaged due to lack of data for an hourly driving pattern throughout the year.
- Regenerative braking energy goes into the on-board electricity storage and is assumed proportional to kilometres driven. As we have no data for braking intensity during the different trips, this seems like a reasonable assumption.
- The energy consumption in the vehicle is divided into consumption by accessory loads and consumption used to propel the vehicle. The former is assumed to receive electricity from the power bus, whereas the propulsion power is delivered from an electric motor and/or an engine, depending on the type of propulsion system. Both the energy consumption of accessory loads and the propulsion power are assumed to be proportional to kilometres driven in each time step.
- An average inverter loss is allocated to all power flows involving AC/DC and DC/AC conversion.
- PHEVs and FCEVs are assumed to use the electric motor until storage is depleted, due to the rather high price difference between fuel and battery use, and the efficiency difference between use of engine and motor. Furthermore, this assumption seems reasonable since batteries developed today already seem to have no loss of effect before almost depleted. The depth of discharge in the batteries is far from the point where the batteries experience any loss of effect. Therefore, the EDVs will be able to accelerate and drive on battery only until switching to engine power.
- Because the model is an investment model, it is a challenge to introduce all the necessary decision variables for an optimization model to be correct, yet still solvable within a reasonable time horizon. Thus, the state of charge of the battery when the vehicle leaves the grid is assumed to be fixed. If not fixed, the model will become nonlinear. Hence, the EDVs leave the grid with a vehicle-dependent but fixed average storage level, given by the load factor. The load factor can, however, be made variable if investments are excluded. This will be commented upon in Section 4.4.5. The use of electric motor versus engine could also be considered for optimization endogenously in the model and is a subject for future research.

4.4.2 Costs

Costs of transport are added to the objective function in Balmorel. Transportation costs include investment costs, operation and management costs, fuel costs and costs of CO_2 emissions. Investment costs $C_{a,v,t}^{\text{invveh}}$ and operation and management costs $C_{a,v,t}^{\text{OMveh}}$ can be calculated identically for all vehicle types (4.1), whereas calculation of fuel costs and CO_2 costs differ between vehicle types. The decision variable $N_{a,v}$ is the number of vehicles of type v in area a:

$$\sum_c \sum_{a \in c, v, t} [(C_{a,v,t}^{\text{invveh}} + C_{a,v,t}^{\text{OMveh}})N_{a,v}] \tag{4.1}$$

Total fuel and CO_2 costs (4.2) for vehicles not connected to the grid depend on fuel costs at each time step t for each vehicle type v in area a, $C_{a,v,t}^{\text{fuelveh}}$, cost of CO_2 emissions in country c, $C_c^{CO_2}$, CO_2 emissions per megawatt-hour $\text{Em}_v^{CO_2}$, number of vehicles, annual driving Dr_v and fuel consumption, $\text{Cons}_v^{\text{fuel}}$:

$$\sum_c \sum_{a \in c, v, t} [(C_{a,v,t}^{\text{fuelveh}} + C_c^{CO_2}\text{Em}_v^{CO_2})N_{a,v}\text{Dr}_v\text{Cons}_v^{\text{fuel}}] \tag{4.2}$$

Total fuel and CO_2 costs for vehicles with grid connection (4.3) depend on the total use of the engine (sum for all vehicles) $O_{a,v,t}^{\text{EnGen}}$, as opposed to the use of the electric motor. η_v^{Eng} is the average engine efficiency:

$$\sum_c \sum_{a \in c, v, t} \left[(C_{a,v,t}^{\text{fuelveh}} + C_c^{CO_2}\text{Em}_v^{CO_2})\frac{O_{a,v,t}^{\text{EnGen}}}{\eta_v^{\text{Eng}}} \right] \tag{4.3}$$

Costs of the FCEVs

The hydrogen add-on for Balmorel was described by Karlsson and Meibom (2008). Activation of the hydrogen add-on makes hydrogen production part of the model. To capture the cost of hydrogen, the hydrogen demand from FCEVs has to be added to existing hydrogen demand as an addition to the hydrogen balance equation. Hydrogen demand for non-plug-ins (4.4) is dependent on the number of vehicles, hydrogen consumption $\text{Cons}_v^{H_2}$ and annual driving:

$$\sum_{a,v,t} (\text{Cons}_v^{H_2} N_{a,v}\text{Dr}_v) \tag{4.4}$$

For plug-ins, hydrogen demand (4.5) is dependent on output from the fuel cell $O_{a,v,t}^{FC}$ and efficiency of the fuel cell η_v^{FC}:

$$\sum_{a,v,t} \left(\frac{O_{a,v,t}^{FC}}{\eta_v^{FC}} \right) \tag{4.5}$$

For electricity and hydrogen, fuel and CO_2 costs are included through increased power or hydrogen demand. Thus, Equations (4.4) and (4.5) correspond to Equations (4.2) and (4.3) without the fuel and CO_2 costs. All costs are added for the total costs of the configuration of the transport system.

4.4.3 Transport Demand

As for electricity and heat, transport demand D_r^{tsp} for each region has to be met by supply. Supply is found by multiplying the number of vehicles, the annual driving per vehicle and UC_v (the utilization of capacity in vehicle technology v), namely the fraction of people transported by each vehicle. Supply is calculated on areas, whereas demand is on regions. Thus, supply areas have to be summed to match the regions:

$$\sum_{a \in r, v} (N_{a,v} t Dr_v UC_v) = D_r^{tsp} \tag{4.6}$$

4.4.4 Power Flows

All constraints except for the demand constraint (4.6) are related to the power flows. Power flows are modelled based on the propulsion systems. To include all types of vehicles, three different propulsion systems are defined:

1. Non-plug-ins
2. BEVs
3. Plug-in series – includes both PHEVs and FCEVs.

For each propulsion system a model of the power flow in the vehicle is constructed. For non-plug-ins, only annual driving and fuel consumption are taken into account, because they do not interact with the power system.

Configuration of the electric and plug-in serials propulsion systems are similar and sketched in Figure 4.1. The ICEs, not using power as propellant, are only to be incorporated in the system

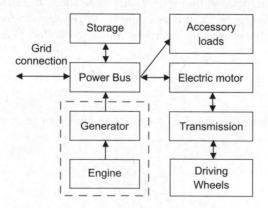

Figure 4.1 Propulsion system configuration of (series) EDVs.

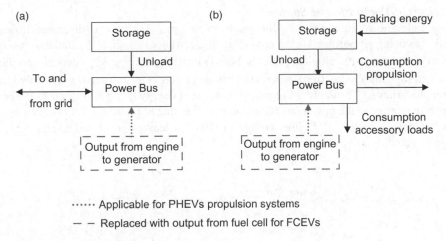

······ Applicable for PHEVs propulsion systems

− − Replaced with output from fuel cell for FCEVs

Figure 4.2 Power flow model of (series) EDVs for (a) vehicles plugged in and (b) vehicles not plugged in.

as a cost. The figure shows the interaction between different units in the vehicle, including grid connection. Power can go both ways from driving wheels to storage and back, and from storage to power grid and back. Power returning from the driving wheels is the regenerated braking energy. The power both ways between storage and the power grid resembles the ability to both load power to the vehicle from the grid (G2V) and unload power from the vehicle to the grid (V2G).

Division of the vehicles into subsystems is needed for modelling the driving and interactions between the power and road transport systems. At the very least, division into storage, engine/fuel cell and the remainder of the system is needed. Further division enables us to study the consequences of improving specific subsystems.

Based on the propulsion system configuration, power flows are sketched in Figure 4.2. The EDVs will be connected to the grid when charging, and the flow of power both when plugged in and when not plugged in has to be modelled. The power flow model reflects the assumption that regenerated braking energy goes into the on-board storage, modelled through the driving efficiency. Only subsystems with more than one ingoing or outgoing power flow are shown. Subsystems with only one ingoing and outgoing power flow (e.g. the electric motor) just calls for a scaling by the efficiency of the subsystem.

Relevant for the power system is the available electricity storage from EDVs at each time step. This is based on storage leaving and arriving with different vehicles, and is captured through the power flow model of the vehicles plugged in (Figure 4.2a). The 'output from engine to generator' for vehicles plugged in resembles the idea that the vehicles can produce power to the grid by use of the engine or fuel cell. For optimizing the use of an electric motor versus the use of fuel cell, gasoline or diesel engine in FCEVs and PHEVs while driving, it is assumed that the electric motor is used until depletion of storage. This assumption is supported by electricity being a cheaper fuel than diesel, gasoline and hydrogen. Therefore, the power flow model for vehicles not plugged in (Figure 4.2b), is based on storage being depleted before use of the engine.

Balancing On-Board Electric Storage

On-board electricity storage can be charged from the grid. The charge–discharge losses are modelled as being proportional to the unloading of electricity storage $S_{a,v,t}^{\text{Unld}}$, and thus modelled as a charge and discharge efficiency η_v^S. On-board electricity storage $S_{a,v,t}^{\text{PI}}$ depends on the last period's storage, the energy from the grid (decision variable) $\text{Gr}_{a,v,t}^{\text{fr}}$ (needs to be scaled by the inverter efficiency η_v^{inverter} in order to equal the amount of energy going into storage), the energy going from storage to the power bus (decision variable, due to possibility of discharging), the charge–discharge losses, $S_{a,v,t}^{\text{Leav}}$ (the storage in vehicles leaving in period t) and $S_{a,v,t}^{\text{Arr}}$ (the storage in vehicles arriving in period t):

$$S_{a,v,t}^{\text{PI}} = S_{a,v,t-1}^{\text{PI}} + \text{Gr}_{a,v,t}^{\text{fr}}\eta_v^{\text{inverter}} - \frac{S_{a,v,t}^{\text{Unld}}}{\eta_v^S} - S_{a,v,t}^{\text{Leav}} + S_{a,v,t}^{\text{Arr}} \quad \forall a \in A; \, v \in V; \, t \in T \quad (4.7)$$

Calculation of storage in vehicles leaving in period t is based on the assumption that all vehicles bring along an average level of storage, given by a percentage of the battery capacity LF_v, the load factor. Furthermore, in accordance with statistical data on transport habits (DTU Transport, n.d.), it is assumed that all vehicles will be parked within a time horizon of 11 h after leaving. γ_v^S is the storage capacity and $\text{PP}_{a,v,i,j}$ is the plug-in pattern for vehicle type v in area a. The plug-in pattern is defined as the percentage of vehicles leaving in hour i and returning to the power grid in hour j.

$$S_{a,v,t}^{\text{Leav}} = \sum_{j=t}^{t+11} \text{PP}_{a,v,t,j}\text{LF}_v\gamma_v^S N_{a,v} \quad \forall a \in A; \, v \in V; \, t \in T \quad (4.8)$$

Storage level in vehicles that arrive in period t depends on the storage in the vehicles when leaving, and thus the capacity of the battery, energy use for driving E_v^{Dr} and energy from braking E_v^{Brk}. The latter two, of course, depend on the distance driven, given as $\text{DD}_{v,1}$ for all full hours of driving, and $\text{DD}_{v,0}$ for the hour in which the vehicles return. A maximization function is used, recognizing that the storage will never be negative:

$$S_{a,v,t}^{\text{Arr}} = \sum_{i=t-11}^{t} \max\left\{\text{PP}_{a,v,i,t}\left(\text{LF}_v\gamma_v^S - [(t-i)\text{DD}_{v,1} + \text{DD}_{v,0}]\left(\frac{E_v^{\text{Dr}}}{\eta_v^S} - E_v^{\text{Brk}}\right)\right); 0\right\} N_{a,v}$$
$$\forall a \in A; \, v \in V; \, t \in T \quad (4.9)$$

The parameter E_v^{Dr} is determined by consumption for propulsion $\text{Cons}_v^{\text{EPrp}}$, accessory loads $\text{Cons}_v^{\text{EAcc}}$, and motor and transmission efficiencies η_v^{mot} and η_v^{trans} respectively:

$$E_v^{\text{Dr}} = \text{Cons}_v^{\text{EAcc}} + \frac{\text{Cons}_v^{\text{EPrp}}}{\eta_v^{\text{mot}}\eta_v^{\text{trans}}} \quad \forall v \in V \quad (4.10)$$

The parameter E_v^{Brk} depends on the regenerated energy going from braking to storage RE_v^{Brk}, as well as on the motor, power bus and transmission efficiencies η_v^{mot}, η_v^{PB} and η_v^{trans} respectively:

$$E_v^{\text{Brk}} = \text{RE}_v^{\text{Brk}} \eta_v^{\text{mot}} \eta_v^{\text{PB}} \eta_v^{\text{trans}} \qquad \forall v \in V \qquad (4.11)$$

Balancing of the Power Bus

Power going out of the power bus needs to equal power going into the power bus at all times. For vehicles plugged in, power from the power bus only goes to the grid, $\text{Gr}_{a,v,t}^{\text{to}}$. Power into the power bus comes from either the engine, $O_{a,v,t}^{\text{EnGenPI}}$, or the on-board storage, $S_{a,v,t}^{\text{Unld}}$:

$$\text{Gr}_{a,v,t}^{\text{to}} = (O_{a,v,t}^{\text{EnGenPI}} \eta_v^{\text{gen}} + S_{a,v,t}^{\text{Unld}}) \eta_v^{\text{PB}} \qquad \forall a \in A; \, v \in V; \, t \in T \qquad (4.12)$$

where $O_{a,v,t}^{\text{EnGenPI}} = 0$ for BEVs and $O_{a,v,t}^{\text{EnGenPI}} = O_{a,v,t}^{\text{FCPI}}$ for FCEVs. The equation includes the possibility of parked vehicles to produce power through use of the engine while parked.

Output from Engine to Generator

Calculation of fuel and CO_2 consumption due to the use of engine power at each time period needs to be kept track of. Output from engine to generator for vehicles plugged in is calculated through Equation (4.12). When assuming that the vehicles deplete the battery before turning on the engine, the calculation of the output from engine to generator for vehicles not plugged in is a question of finding the time step when the vehicles start using the engine. In Figure 4.3, the area above the x-axis resembles use of on-board storage and the area below the x-axis resembles use of engine. To find the crossing of the x-axis, we need to distinguish between three operating situations: the vehicle returns to the grid before the storage is depleted (C), the vehicle returns in the same time period as the storage is depleted (B) and the vehicle returns in time periods after the storage is depleted (A). The first case does not involve usage of the engine and, therefore, is not treated in the following.

The distance driven until storage is depleted will be

$$\text{DD}_v^{\text{deplete}} = \begin{cases} (t_c - i)\text{DD}_{v,1} + \text{DD}_{v,0} & \text{if } t_c = j \\ (t_c - i + 1)\text{DD}_{v,1} & \text{if } t_c < j \end{cases}$$

The vehicle leaves the power grid in time period $i = 1, 2, \ldots, t$ and returns to the power grid in time period $j = i, i + 1, \ldots, i + 11$. To find the time period where the on-board storage is depleted t_c, the following is calculated.

If $t_c = j$, t_c is found by setting the capacity left on the battery equal to zero. Calculation of the capacity used includes both full hours of driving and the driving in the hour of return:

$$\text{LF}_v \gamma_v^S - [(t_c - i)\text{DD}_{v,1} + \text{DD}_{v,0}] \left(\frac{E_v^{\text{Dr}}}{\eta_v^S} - E_v^{\text{Brk}} \right) = 0 \quad \forall v \in V; \, t \in T \qquad (4.13)$$

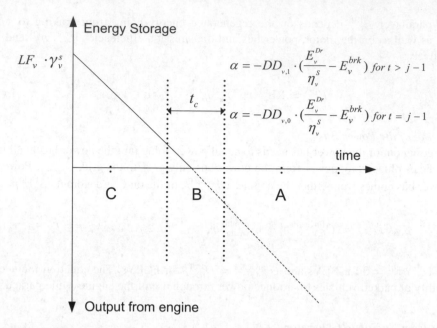

Figure 4.3 The use of energy storage versus engine depending on time. α is the slope of the line and $\text{LF}_v\gamma_v^S$ the storage level when the vehicle is leaving the grid. t_c is the time period where the on-board storage is depleted.

If $t_c < j$, t_c is found in the same way as above except that the capacity used is calculated using only full hours of driving:

$$\text{LF}_v\gamma_v^S - (t_c - i + 1)\text{DD}_{v,1}\left(\frac{E_v^{\text{Dr}}}{\eta_v^S} - E_v^{\text{Brk}}\right) = 0 \qquad \forall v \in V;\ t \in T \qquad (4.14)$$

The term $t_c - i$ indicates the number of hours before the storage is depleted, and thus the vehicles start using the engine. The index $t_c - i$ can be found using Equations (4.13) and (4.14).

The index $t_c - i$ can be calculated for each vehicle type, since all the other parameters are fixed on vehicle type level. Output from engine to generator can then be calculated for all combinations of vehicles that leave in time period i and return in time period j. $\sum_{i=1}^{t}\sum_{j=t}^{t+11}\text{PP}_{a,v,i,j}$ is the total share of all vehicles not plugged in at time t in area a. The calculation of the output from engine to generator in time period t now depends on t being smaller than, equal to or greater than t_c. In situation A, where electric storage is depleted (Figure 4.3), the engine output in each hour of driving will be

$$O_{a,v,t>t_c}^{\text{EnGenNPI}} = \sum_{i=1}^{t}\sum_{j=t}^{t+11} N_{a,v}\text{PP}_{a,v,i,j}\text{DD}_{v,1}\left(\frac{E_v^{\text{Dr}}}{\eta_v^S} - E_v^{\text{Brk}}\right) \qquad \forall v \in V;\ t \in \{t' \in T : t' > t_c\}$$

$$(4.15)$$

If in situation B, the electric storage depletes in time period t_c (Figure 4.3). The output from engine to generator is

$$O_{a,v,t=t_c}^{\text{EnGenNPI}} = -N_{a,v} \sum_{i=1}^{t} \sum_{j=t}^{t+11} \left\{ (PP_{a,v,i,j} \left[LF_v \gamma_v^S - (t_c - i + 1)DD_{v,1} \left(\frac{E_v^{\text{Dr}}}{\eta_v^S} - E_v^{\text{Brk}} \right) \right] \right\}$$

$$\forall a \in A; \, v \in V; \, t \in \{t' \in T : t' = t_c\} \tag{4.16}$$

In Equations (4.15) and (4.16), $DD_{v,0}$ is included if the vehicle returns in the time period under consideration; that is, $j = t$. Thus, $(t_c - i + 1)DD_{v,1}$ is replaced with $(t_c - i)DD_{v,1} + DD_{v,0}$.

Finally, in situation C, the vehicle only uses electric storage, such that the sum of the results of Equations (4.15) and (4.16) gives the total output from engine to generator in period t for vehicles not plugged in. The total output from engine to generator is

$$O_{a,v,t}^{\text{EnGen}} = O_{a,v,t>t_c}^{\text{EnGenNPI}} + O_{a,v,t=t_c}^{\text{EnGenNPI}} + O_{a,v,t}^{\text{EnGenPI}} \quad \forall a \in A; \, v \in V; \, t \in T \tag{4.17}$$

As with vehicles plugged in, $O_{a,v,t}^{\text{EnGenNPI}} = 0$ for BEVs. $O_{a,v,t}^{\text{EnGenNPI}} = O_{a,v,t}^{\text{FCNPI}}$ and $O_{a,v,t}^{\text{EnGen}} = O_{a,v,t}^{\text{FC}}$ for FCEVs.

Storage Level

The storage level is to stay between zero and aggregated storage available.

$$0 \leq S_{a,v,t}^{\text{PI}} \leq N_{a,v} \left(1 - \sum_{i=1}^{t} \sum_{j=t}^{t+11} PP_{a,v,i,j} \right) \gamma_v^S \quad \forall a \in A; \, v \in V; \, t \in T \tag{4.18}$$

Capacity Restrictions on Loading and Unloading of On-Board Storage, Power Flow to and From Grid, and Engine Output

These restrictions depend on the single vehicle capacities of respectively loading γ_v^{Sld}, unloading, grid connection and engine output multiplied by the number of vehicles plugged in at each time step. As an example, the power flow into storage when plugged in at each time step is given by

$$\text{Gr}_{a,v,t}^{\text{fr}} \eta_v^{\text{inverter}} \leq N_{a,v} \left(1 - \sum_{i=1}^{t} \sum_{j=t}^{t+11} PP_{a,v,i,j} \right) \gamma_v^{\text{Sld}} \quad \forall a \in A; \, v \in V; \, t \in T \tag{4.19}$$

Similar restrictions apply for unloading of on-board storage, power to and from grid, and engine output, although not shown here.

Addition to the Electricity Flow Balance Equation in Balmorel

For balancing the power flows in the power system, the net power flow from the transport system to the power system is added to the electricity balance restriction in Balmorel.

$$+ \sum_{a \in r, v} (\mathrm{Gr}^{to}_{a,v,t} - \mathrm{Gr}^{fr}_{a,v,t}) \qquad \forall t \in T \tag{4.20}$$

4.4.5 Variable Load Factor

As mentioned, the investment model does not allow for vehicles to leave with an endogenously optimized battery load. However, by excluding the investment option, results can be obtained with optimized battery load. In order to make the load factor in the model vary over time, the variable has to be time dependent. Thus, LF_v will be changed to $\mathrm{LF}_{v,t}$. Furthermore, the vehicle groups need to be further specified; thus, the driving pattern given in Section 4.5.2 needs to be divided into groups as done, for example, in Kristoffersen *et al.* (2011) in order to get a more valid analysis.

4.4.6 BEVs

BEVs have a shorter driving range than the other vehicle types owing to battery size. Thus, different driving patterns, plug-in patterns etc. are needed. In order to ensure that all trips are covered, the other vehicle types are to cover the longer trips. Therefore, the final driving patterns are not known until the share of BEVs is known. This makes the inclusion of BEVs an iterative process.

Owing to costly calculation time, it is chosen to let the model invest in BEVs using the same driving patterns as the remaining vehicles. In case it is optimal to invest in BEVs, the driving patterns and plug-in patterns for these will be introduced accordingly and the model will be rerun.

4.4.7 EDVs Contributing to Capacity Credit Equation

To make sure the power system is configured to meet demand at all times, a capacity credit equation is introduced. This equation ensures enough power generating capacity in the system to meet peak-load demand (taken from a '10-year' winter); that is, the largest demand during any 24 h period. Peak demand is usually experienced in the morning when people get up and in the evening when people come home from work and start cooking. In the existing Balmorel model, it is assumed that 99% of all dispatchable power plant capacities in area a at time t ($\gamma^{dispatch}_{a,t}$) are available at peak load, while 14% of wind power capacity ($\gamma^{wind}_{a,t}$) is assumed to contribute to meet peak-load demand. The capacity credit equation for each area a is

$$\sum_{a \in c, p} (0.99 \gamma^{dispatch}_{a,p,t} + 0.14 \gamma^{wind}_{a,p,t}) \geq D^{peakload}_{c,t} \qquad \forall c \in C; \; t \in T \tag{4.21}$$

One could argue that the EDVs could help providing capacity in peak load hours contributing to the capacity credit equation. Comparing driving patterns with the peak load hours, clearly most vehicles are parked in peak load hours. Assuming that the EDVs are plugged in when parked, the vehicles are available to the power system – also during peak hours.

In order to incorporate the vehicles in the capacity credit equation, an estimate of the available battery capacity is needed. Some vehicles will be leaving the grid soon; therefore, they experience restrictions on the on-board storage. Furthermore, some batteries have just been used, and only some or none of the battery capacity is left. Hence, a limited amount of energy is available for peak load hours. As shown by Letendre and Kempton (2002), at least 92% of the vehicles are parked at all times. Conservatively, an estimation can be that only 70% of the PHEVs (corresponding to 3/4 of the 92%) and 76% of BEVs (corresponding to 5/6 of the 92%) are available for the electricity system during peak load. The larger share estimated for the BEVs is due to the shorter driving ranges and, thus, a larger share of these are parked at all times.

To find out how much the EDVs can unload and, thus, with how much they can contribute to the capacity credit equation, the storage level when the peak load starts needs to be estimated. Assuming that the storage is at an average level of 50% by the time peak-load hours start, an assumption of an average of 30% of the battery available for peak load seems to be conservative. Based on these estimates, the addition from the road transport to the capacity credit equation is

$$0.3 \sum_{a \in c} (N_{a,\text{BEV}} \times 0.76 \gamma^S_{\text{BEV}} + N_{a,\text{PHEV}} \times 0.7 \gamma^S_{\text{PHEV}}) \qquad (4.22)$$

This is only for a short period of time, though, since the battery's state of charge will be decreasing and will reach the minimum state of charge.

The grid capacity also puts a limit on the discharge. Each vehicle has an assumed grid connection capacity γ^{Gr} of 6.9 kW, corresponding to a standard 230 V connection with three phases of 10 A each. Hence, the storage in the BEVs should last, but the storage for the PHEVs will not. Scaling the storage in the PHEV with 2/3 (an estimate, trying to ensure that storage will last) changes the contribution from the EDVs to

$$\sum_{a \in c} \left(N_{a,\text{BEV}} \times 0.76 + N_{a,\text{PHEV}} \times 0.7 \times \frac{2}{3} \right) \gamma^{\text{Gr}} \qquad \forall a \in A \qquad (4.23)$$

We now undertake a calculation to verify the availability of capacity. Take a vehicle fleet with 1 million pluggable EDVs and assume the on-board storage of these plug-ins is half full. It then takes more than 3 h to empty the on-board storage, leaving respectively 5% and 20% of on-board storage in the PHEVs and BEVs. Peak load usually lasts for a period of 1–2 h; thus, even with vehicles leaving the grid with more storage than those arriving, the storage should last.

The capacity credit equation including EDVs then is

$$D^{\text{peakload}}_{c,t} \leq \gamma^{\text{Gr}} \sum_{a \in c} \left(N_{a,\text{BEV}} \times 0.76 + N_{a,\text{PHEV}} \times 0.7 \times \frac{2}{3} \right) + \sum_{a \in c, p} (0.99 \gamma^{\text{dispatch}}_{a,p,t} + 0.14 \gamma^{\text{wind}}_{a,p,t})$$

$$\forall c \in C; \ t \in T \qquad (4.24)$$

4.5   Case Study

To illustrate the benefits of integrating the power and road transport systems, the model is applied to a northern European case for the year 2030. Different scenarios are introduced and the consequences of the optimal configuration and operation of the power and transport systems are analysed.

In order to be able to run the model within reasonable computation time, Norway, Sweden and Finland represent one region each. Germany is divided into two regions in order to represent the transmission bottlenecks between northern Germany with a large share of wind power and the consumption centres in central Germany. Denmark is split into two regions representing western Denmark being synchronous with the UCTE power system and eastern Denmark being synchronous with the Nordel power system.

A number of inputs are required for running the model; for example, fuel prices, CO_2 prices, demand data, and vehicle and power system technology data, which are all given exogenously. In the base case, oil prices are assumed to be \$120/barrel, assuming constant price elasticities as in Karlsson and Meibom (2008). CO_2 prices are assumed to be 40 €/ton. Data, such as the demand data, are to be given on a regional level (Table 4.1). Currently, the transmission capacity between the two regions in Denmark is at 600 MW, but for the year 2030 a transmission capacity of 1.2 GW and a transmission loss of 1% are assumed.

In order for Balmorel to balance power supply and demand, new technologies must be available for investment. Table 4.2 shows the technologies available for investment in 2030 in the base case. With the analysis focusing on competition between technologies and the integration of more renewables, this list of technologies is thought of as a good basis.

As for the power system, balancing the transport supply and demand requires investment opportunities in different technologies. With focus on incorporation of more renewable energy sources through the introduction of different EDVs, it is decided to consider four different vehicles technologies: diesel ICE, series PHEV (diesel), series plug-in FCEV (referred to as FCEV) and BEV. The four technologies compete both in costs and in delivering benefits for the power system.

4.5.1   Vehicle Technologies

The vehicles can be divided into two categories: ICEs and EDVs, with the latter covering all vehicles that use batteries for driving, both hybrids and all electrics. From a

Table 4.1 Demand input data year 2030 (for EU countries the source is European Commission, Directorate-General for Energy and Transport (2010) – for Norway, Swedish demand has been scaled according to the situation today).

Demand	Denmark east	Denmark west	Sweden	Norway	Finland	Germany
Electricity (TWh/yr)	16	24	153	133	104	620
District heat (TWh/yr)	12	15	46	3	56	102
Transport (10^9 persons km/yr)	31	41	148	69	86	1262

Table 4.2 Technology investment options in the simulation. Investment costs for heat storage and hydrogen storage are given as 10^6 €/MWh storage capacity.

Technology	Source	Inv. costs[a] (10^6 €/MW)	V O & M cost (€/MWh)	F O & M cost (10^3 €/(MW yr))	Efficiency[b]
Onshore wind	Danish Energy Authority (2010)	1.22	11.5	—	1
Offshore wind	Danish Energy Authority (2010)	2.2	15	—	1
Coal extraction, steam turbine	Danish Energy Authority (2010)	1.1	—	34	0.51
Open-cycle gas turbine	Danish Energy Authority (2010)	0.57	3	8.6	0.42
Combined-cycle gas turbine, condensing	Delucchi et al. (2000)	0.56	3.4	21.4	0.58
Combined-cycle gas turbine, extraction	Danish Energy Authority (2010)	0.47	4.2	—	0.61
CHP plant, biomass (medium)	Danish Energy Authority (2010)	1.6	3.2	23	0.485
CHP plant, biomass (small)	Danish Energy Authority (2010)	4	—	140	0.25
Nuclear	Delucchi et al. (2000)	2.81	7.7	55.5	0.37
Heat storage	Ellehauge and Pedersen (2007)	0.00178	—	—	0.99
Heat pump	Danish Energy Authority (2010)	0.55	—	3	3
Electric boiler	Danish Energy Authority (2010)	0.06	0.5	1	0.99
Heat boiler, biomass	Danish Energy Authority (2010)	0.5	—	23.5	1.08
Heat boiler, natural gas	Danish Energy Authority (2010)	0.09	—	0.32	1

Fuel column:
Onshore wind: —; Offshore wind: —; Coal extraction, steam turbine: Coal; Open-cycle gas turbine: Natural gas; Combined-cycle gas turbine, condensing: Natural gas; Combined-cycle gas turbine, extraction: Natural gas; CHP plant, biomass (medium): Wood; CHP plant, biomass (small): Wood-waste; Nuclear: Uranium; Heat storage: —; Heat pump: Electric; Electric boiler: Electric; Heat boiler, biomass: Wood; Heat boiler, natural gas: Natural gas.

[a] Investment costs will be annualized with a discount rate of 3% (low rate is due to fixed prices).
[b] For heat pumps coefficient of performance (COP).

Table 4.3   Vehicle technology investment options (Nguyen and Stridbaek, 2007).

Vehicle type	Annualized inv. costs (€)	O & M costs (€ /year)	Electric storage capacity (kWh)
ICE	1065	1168	0.8
BEV	1513	1101	50
PHEV	1484	1168	10
FCEV	1893	1101	10

modelling perspective the interesting vehicles are those that can be plugged in (PHEVs or BEVs).

The model developed for Balmorel supports different vehicle technologies, including:

- **ICEs:** These will have no influence on the power system, only on fuel and CO_2 costs. Diesel as propellant has been chosen because of the fuel economy being better than that of the gasoline vehicles.
- **PHEVs:** These influence the power system as they can be charged from the power grid and might also be able to discharge (V2G). PHEVs can in principle deliver power to the grid with the motor running as well, although the efficiency is low and it is not environmentally friendly and, thus, the sustainability is debatable.
- **BEVs:** These influence the power system with charge and possibly discharge.
- **FCEVs:** These are like PHEVs with a fuel cell running on; for example, hydrogen as a range extender instead of an engine. In this situation, sustainability can be maintained with the fuel cell running to provide back-up power for the electricity grid. The low efficiency of this solution is up for discussion.

The cost and electric storage capacity for the four vehicle technologies included in the base case are given in Table 4.3. The size of the electric storage capacity, shown in the table, reflects the usable size of the battery. Today, the electric vehicle efficiency is approximately 5 km/kWh (Delucchi *et al.*, 2000; Denholm and Short, 2006; Short and Denholm, 2006), leading us to believe that the efficiency will reach approximately 7 km/kWh by 2030. The battery size for BEVs by 2030 is believed to provide a driving range of approximately 350 km. For PHEVs the batteries could be almost as large as for BEVs, but the desire to avoid the additional weight, as well as the trade-off between additional driving range and additional cost, leads us to believe that a battery covering a driving range of approximately 65 km is reasonable for everyday purposes.

FCEVs have been considered. However, the price by 2030 is still too high for investments to be made in these vehicles. Thus, these are excluded in the remainder of the case study.

4.5.2   *Driving Patterns/Plug-in Patterns*

Each vehicle type is associated with a particular plug-in pattern extracted from historical driving patterns (DTU Transport, n.d.). A plug-in pattern consists of percentages for each time step, each pair representing the fraction of vehicles leaving at the particular time step i and returning in time step j, thus returning in $j - i$ future time steps. A percentage is given for each combination of leaves and arrivals within a 24 h time horizon. All EDVs are assumed to be plugged in when not driving.

Converting Data from Danish Transport Habits into Data for Balmorel

The survey of Danish transport habits was made in 2006 (DTU Transport, n.d.). The transport habits consist of a number of data related to trips and transport, including walking, bicycling and driving a vehicle. The following vehicle data are found relevant for our analysis:

- **Start time for the trip:** The time of day on an hourly basis.
- **Time spent on the trip:** Minutes split into intervals. In the survey, the first hour is split into six intervals, second hour into two intervals, third hour is one interval, fourth and fifth hours are one interval, sixth through tenth hours one interval and, finally, eleventh hour and above are one interval. As Balmorel uses an hourly time scale, the survey data have been adjusted accordingly.
- **Whether driving the vehicle or a passenger:** We consider only trips by the driver and ignore passenger information.
- **Amount of transport per weekday:** An average of kilometres driven per person on the particular weekday.

Based on these data the driving patterns are extracted (see Table 4.4). Because Balmorel uses hourly time steps, the data are changed accordingly. It is easy with the first three hours.

Table 4.4 Conversion of length of trip from the Danish transport survey (based on minutes – 'survey minutes' column) to Balmorel (based on hours).

Balmorel hour	Algorithm estimate	Survey minutes
		1–5
		6–10
	sum	11–20
1	of	21–30
	all	31–45
		46–60
	sum	61–90
2	of all	91–120
3	equal	121–180
4	2/3	
5	1/3	181–300
6	1/3	
7	4/15	
8	1/5	301–600
9	2/15	
10	1/15	
11	equal	≥601

Table 4.5 Driving patterns transformed into trips by the hour. Percentage of vehicles coming home after 1–11 h, for each hour in a 24 h time interval.

Return time											
Time leaving	1	2	3	4	5	6	7	8	9	10	11
1	85%	9%	2%	2%	1%	0%	0%	0%	0%	0%	0%
2	82%	5%	9%	2%	1%	0%	0%	0%	0%	0%	0%
3	66%	11%	0%	0%	0%	8%	6%	5%	3%	2%	0%
4	71%	12%	3%	9%	5%	0%	0%	0%	0%	0%	0%
5	71%	21%	2%	3%	1%	1%	1%	0%	0%	0%	0%
6	79%	12%	5%	2%	1%	1%	0%	0%	0%	0%	0%
7	83%	12%	3%	1%	1%	0%	0%	0%	0%	0%	0%
8	84%	10%	3%	1%	1%	0%	0%	0%	0%	0%	0%
9	78%	13%	4%	2%	1%	1%	1%	0%	0%	0%	0%
10	76%	13%	5%	4%	2%	0%	0%	0%	0%	0%	0%
11	78%	11%	4%	3%	1%	1%	1%	0%	0%	0%	0%
12	80%	10%	4%	3%	1%	1%	1%	1%	0%	0%	0%
13	79%	11%	5%	2%	1%	0%	0%	0%	0%	0%	0%
14	82%	10%	4%	2%	1%	0%	0%	0%	0%	0%	0%
15	83%	11%	3%	2%	1%	0%	0%	0%	0%	0%	0%
16	85%	10%	3%	1%	1%	0%	0%	0%	0%	0%	0%
17	84%	10%	3%	1%	1%	0%	0%	0%	0%	0%	0%
18	86%	9%	2%	1%	1%	0%	0%	0%	0%	0%	0%
19	86%	9%	2%	2%	1%	0%	0%	0%	0%	0%	0%
20	85%	12%	2%	1%	1%	0%	0%	0%	0%	0%	0%
21	85%	11%	2%	1%	1%	0%	0%	0%	0%	0%	0%
22	86%	9%	2%	1%	1%	0%	0%	0%	0%	0%	0%
23	88%	9%	1%	0%	0%	1%	1%	0%	0%	0%	0%
24	88%	7%	4%	0%	0%	0%	0%	0%	0%	0%	0%

For the fourth and fifth hours an approximation is made, recognizing that the number of trips decreases with increasing trip length. Two-thirds of the trips in the interval are assigned to the fourth hour, whereas one-third of the trips are assigned to the fifth hour. The same challenge is seen for the next interval. The last interval contains very few observations; thus, all of these vehicles are considered to be driving for a maximum of 11 h.

Having changed the intervals, the number of vehicles leaving in each hour are split into these intervals; thus, they are split into how many hours the trip takes (see Table 4.5). These numbers are then changed to percentages of vehicles leaving at a particular hour (e.g. at 7 o'clock) and then returning after 1–11 h, which sums up to 100% for each hour. Based on the same data, the share of vehicles leaving in each hour has been found. Thus, multiplying the two gives percentages analogous to the driving pattern for a 24 h time interval, but without information of the difference between weekdays.

Based on the investigation of transport habits, weights for each weekday are found. Multiplying the 24 h driving pattern with the weights gives a rather detailed weekly driving pattern. It could be argued that the driving pattern should be seasonal, based on feasts, vacations etc. However, the data are not that detailed and average weekly data should give us a good picture of the interaction between the power system and the road transport system. Distinguishing between driving on the different weekdays has been done according to the transport survey.

It is assumed that each vehicle has an approximate driving distance of 100 km in full hours of driving and 36 km in hours when they return to the grid. Thus, a 2 h trip covers a distance of 136 km.

For BEVs the driving patterns are somewhat more uncertain. First of all, the BEVs cannot drive very far; thus, either other vehicles are needed for the longer trips or our driving habits will have to change. Both things will probably happen. However, in order to analyse BEVs compared with PHEVs, the number of vehicles being second vehicle, third vehicle etc. in a family has been found using Danish Statistics (2009). For some analyses it has been assumed that these vehicles are BEVs. The vehicles amount to 25% of all the vehicles in Denmark. Driving patterns are then adjusted accordingly, in order to only include the first 3 h of driving (no matter what time the vehicle leaves) for the BEVs, using the 350 km driving range restriction on the battery. Hence, the driving patterns for the remaining vehicles have to be adjusted likewise. In order to compensate for the long trips not driven by the BEVs, a larger share of the remaining vehicles are driving the long trips.

Assuming that all vehicles are plugged in when parked, the plug-in pattern is derived directly from the driving pattern. However, it is possible to use other plug-in patterns.

4.6 Scenarios

The focus is on the investigation of an optimal configuration of the power system depending on flexibility of the vehicles. What is the influence of an introduction of V2G and what are the consequences/benefits of an introduction of a percentage of BEVs? It is believed that V2G has some influence on the configuration of the power system, since being able to deliver power back to the grid delivers a greater flexibility than just flexible demand. This could very well mean introduction of more wind in the cases with V2G facilities available.

In order to investigate the above, a number of scenarios have been set up. First of all, the base case, described in Section 4.5, will be run and the results analysed. Based on this, the following will be changed first separately and subsequently some of them simultaneously, creating many different scenarios (Table 4.6):

- The inclusion of the G2V and V2G facility.
- Fuel price sensitivities are set to low at $90/barrel and high at $130/barrel.
- CO_2 price sensitivities are set to low at 10 € /ton and high at 50 € /ton.
- Having a BEV as the second vehicle, third vehicle etc. in the house. Danish Statistics (2009) show that 25% of all trips are driven with these vehicles in Denmark; thus, 25% of all trips will be covered by BEVs in this model run.

4.7 Results

The model was run on a computer with a 2.99 GHz processor and 7.8 GB RAM. Calculation time for running 7 weeks with 168 time steps is approximately 31 h.

Table 4.6 Scenarios.

Scenario	V2G	G2V	BEV	Oil	CO_2
Base	+	+			
NoV2G		+			
NoV2GG2V					
BEV	+	+	+		
HighOil	+	+		high	
HighCO$_2$	+	+			high
HighOilCO$_2$	+	+		high	high
LowOil	+	+		low	
LowCO$_2$	+	+			low
LowOilCO$_2$	+	+		low	low

4.7.1 Costs

The total costs of the integrated power and road transport system vary with the different scenarios (Figure 4.4). For the base scenario this amounts to €203×10^9. Forcing BEVs into the system increases the costs the most, due to the battery costs.

No interaction between the power and transport system, of course, results in investments in ICEs. Introducing G2V (the possibility of investing in EDVs) saves €6.2×10^9. Introduction of V2G saves another €18×10^6 (Figure 4.4).

4.7.2 Investments and Production

In seven out of ten scenarios, it is beneficial to invest in PHEVs. In the other three scenarios, all or almost all investments are placed in ICEs. The first scenario is when forced (no interaction

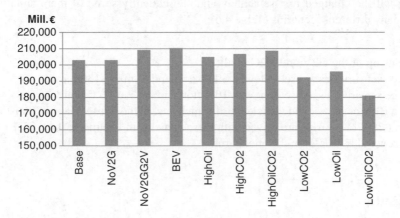

Figure 4.4 Total costs (€10^6) for optimal investments and operation of the integrated power and transport system under the different scenarios.

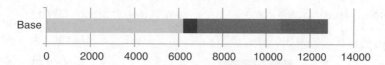

Figure 4.5   Base case yearly power generation based on fuel type, Danish power system.

between the vehicles and the power grid, and thus no possibilities for recharging the EDVs), and the others are the low oil costs scenario and the low oil-and-CO_2 costs scenario.

Power plant investments, on the other hand, vary depending on the scenarios. For Denmark, introducing PHEVs result in higher investments in and generation of wind power (Figure 4.5 and Figure 4.6) – an increase in production of 5.9 TWh. The increase in wind power production more than exceeds the power supply for the PHEVs (5.4 TWh/yr). Thus, in the Danish system the vehicles will be sustainable to the degree that they drive on electricity from wind power.

Focusing on the German power system, investments and generation are in coal, except for the low oil scenario (Figure 4.7 and Figure 4.8). Power production in Germany is characterized by a fixed level of nuclear power on 129 TWh, and varying levels of production on coal and lignite (431 TWh and 150 TWh respectively in the base case). Hence, in Germany the PHEVs are driving on coal and lignite (95.1 TWh/year).

For Norway, investments are primarily in wind power, whereas generation primarily is hydro (Figure 4.9 and Figure 4.10). The investments and electricity production using wind

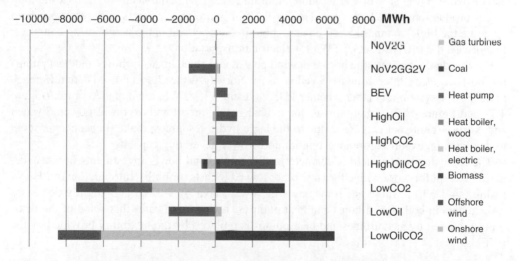

Figure 4.6   Deviations from base case power generation based on fuel types, Danish power system.

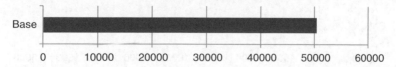

Figure 4.7   Base case yearly power generation based on fuel type, German power system.

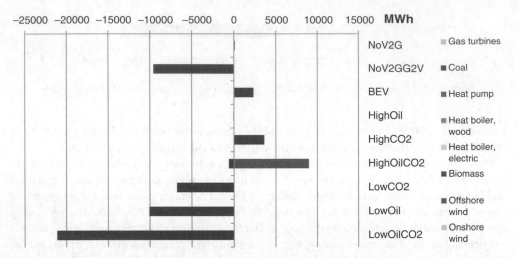

Figure 4.8 Deviations from base case power generation based on fuel types in the German power system.

vary according to the scenarios, due to varying import and export to the neighbouring countries depending on the prices of alternative power generation in these countries. Furthermore, the use of hydro power in Norway is stable at around 128 TWh, equivalent to 72% of the total power production in the base case. The facilities of hydro power plants with reservoirs help integrate the high amount of wind power – both in Norway and in the neighboring countries. In Norway, the PHEVs use 5.2 TWh for electric transportation.

For Sweden, most investments are in wind power, and some in gas turbines and heat pumps for most scenarios. Interesting for Sweden, as for Norway, is the almost nonexistent difference in the power system no matter whether PHEVs, using 11.1 TWh, are included or not (Figure 4.11 and Figure 4.12). Again, this is due to changes in import and export of power. Sweden and Norway being net exporters, due to the large hydro resources, make it cheaper for them to cut down on export before investing in more power generating capacity.

The Finnish power system is characterized by high wind power investments in most scenarios, except for those with either low oil or low CO_2 costs (or both). Introduction of PHEVs (using 6.4 TWh) in this system increases the use of coal by 9 TWh and slightly decreases investments in both gas turbines and heat pumps. The latter indicates that some of the flexibility provided by gas turbines and heat pumps is replaced by the flexibility provided by the PHEVs.

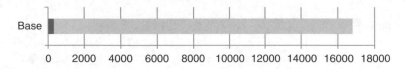

Figure 4.9 Base case yearly power generation based on fuel type, Norwegian power system.

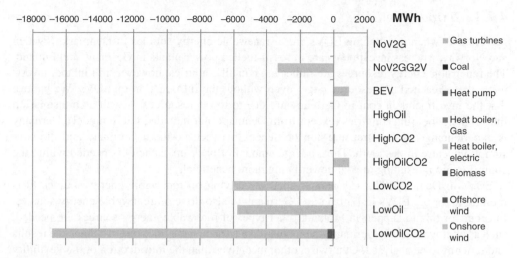

Figure 4.10 Deviations from base case power generation based on fuel types, Norwegian power system.

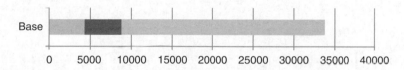

Figure 4.11 Base case yearly power generation based on fuel type, Swedish power system.

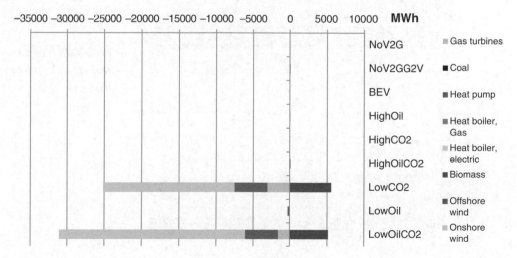

Figure 4.12 Deviations from base case power generation based on fuel types, Swedish power system.

4.7.3   Introducing EDVs

Investments when introducing EDVs are in renewable energy sources in Denmark, Sweden and Norway, whereas investments are in fossil-fuelled power plants in Germany and Finland. The renewable energy resources are rather large in all countries; however, in Finland, investments have reached the resource target even without the introduction of EDVs. We believe that the investments in coal in Germany are due to lower resources as well as transmission lines to neighbouring countries (except from Denmark, not included in the model). Germany is experiencing a power transmission bottleneck between northern Germany and the consumption centre in the south. Thus, a large amount of back-up capacity is needed within the country/region, resulting in coal power being more beneficial.

EDVs in Denmark, Norway and Sweden are driving on renewable energy sources. It is preferable for the EDVs in Finland and Germany to also drive on renewable energy sources. In order for this to happen in Finland, other types of renewable energy sources are needed, such as photovoltaics. In Germany, the inclusion of transmission lines might change the results and remains to be analysed. Otherwise, other incentives than the introduction of the flexibility provided by EDVs are needed.

When excluding EDVs from the power system (no V2G and no G2V), the electricity prices show large fluctuation (Figure 4.13). Introduction of G2V, and thus PHEVs, removes a lot of

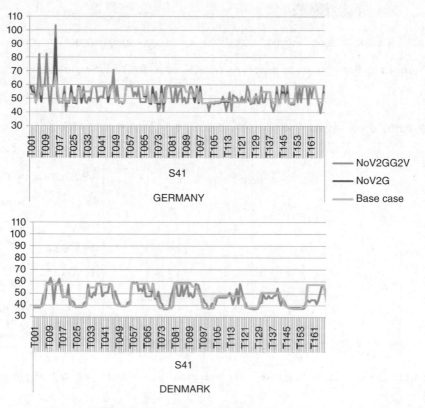

Figure 4.13   Electricity price fluctuations.

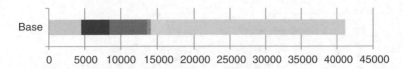

Figure 4.14 Base case yearly power generation based on fuel type, Finnish power system.

the fluctuation in electricity prices, indicating that electricity production is more stable. Thus, the PHEVs supply the flexibility needed in order to meet demand. Furthermore, introduction of V2G cuts off some peaks in the electricity prices in the German power system. It is interesting that the prices of electricity do not increase with the introduction of PHEVs in the energy system (except for Sweden). This is due to better utilization of the base-load power plants being able to produce more constantly. Hence, besides the power system operation being less costly, the PHEVs do provide benefits to the electricity system, no matter whether they make the power system more sustainable or not.

Introduction of 25% BEVs changes investments in the power system as well as in generation. Generally, investments and use of gas turbines decreases. For Denmark and Norway, investments and use of wind power increase (Figure 4.6 and Figure 4.10); for Germany (Figure 4.8) and Finland (Figure 4.14 and Figure 4.15) the increase is in coal steam turbine. Electricity prices smooth out and stay at the same level as without BEVs for all countries but Sweden. Hence, BEVs provide flexibility to the power system, enabling higher wind or coal penetration and more stable production.

4.7.4 Charging the PHEVs

Looking at the use of energy over the day, part of the loading is done during night-time, although there is more charging than expected in the daytime (Figure 4.16). Charging during

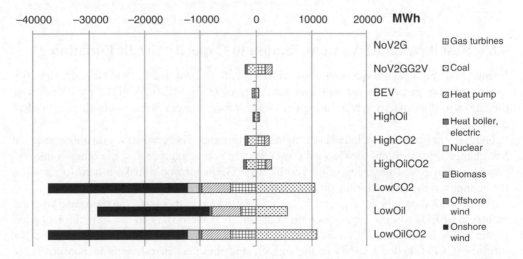

Figure 4.15 Deviations from base case power generation based on fuel types, Finnish power system.

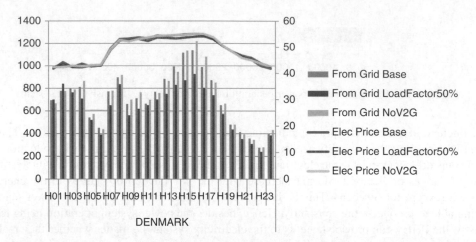

Figure 4.16   Power from grid to vehicles (megawatt-hours) and average electricity prices.

night-time is due to lower electricity prices, again due to a large amount of, otherwise unused, production capacity, including renewable energy sources that might have to be shut down if electricity demand is too low.

The daytime charging is due to the rather strict assumptions about the load factor of the vehicles leaving the grid. The load factor of the PHEVs is set to 100% in order for the vehicles to be able to drive as far as possible on electrical power. If all PHEVs leave the grid with a load factor of 100%, charging during the day is required to meet the restrictions and, thereby, fixed to be rather high. This does not leave much flexibility to the power system to optimize. Reduction of the load factor to 50% does reduce the charging throughout the 24 h period. Furthermore, the reduced load factor increases the total costs of the system by €3.3 × 10^9 because of the increased use of diesel for some vehicles.

4.8   Results from EDVs Contribution to Capacity Credit Equation

In this analysis, EDVs have been allowed to be part of peak-load capacity; thus, they are made an active part of the capacity credit balance equation (4.24). With the contribution of the EDVs to the capacity credit during peak load, power system investments change in most countries.

Germany has not been included in the figures for this analysis because they do not experience any change. Furthermore, the German power system is much larger than the other countries' power systems; thus, inclusion of the results from Germany would make it hard to focus on the changes in the results for the other countries.

Figure 4.17 shows the investments in both the base case without and the base case with the inclusion of EDVs in the capacity credit equation. It is interesting to see how the inclusion of the EDVs in the capacity credit equation removes all investments in combined cycle gas turbines (CCGT, called CC-NG in the figure), and thus less investments in nonrenewable power sources.

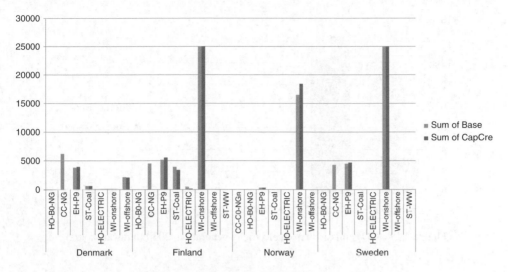

Figure 4.17   Investments in 2030 (megawatts).

Furthermore, there is a slight change in other investments. More wind power investments in Norway and less coal steam turbine investments in Finland are the largest changes.

Even more interesting is the electricity generation shown in Figure 4.18. Here, it becomes evident that the CCGTs, which are no longer invested in, actually hardly provided the power system with any load! Hence, these plants have been there only to satisfy the restrictions on the capacity for peak-load hours.

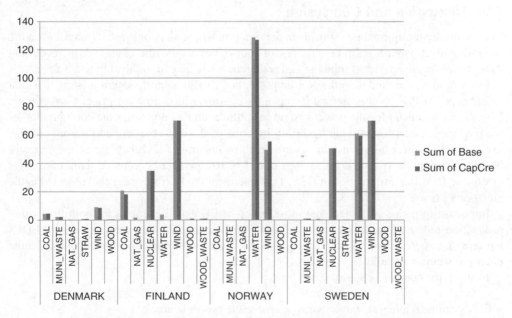

Figure 4.18   Electricity generation on fuel types, 2030 (terawatt-hours).

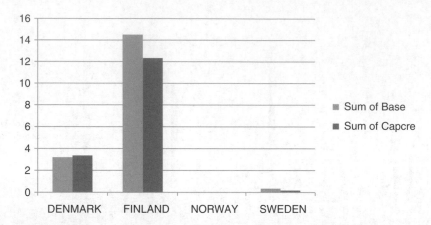

Figure 4.19 CO_2 emissions from electricity generation, 2030 (terawatt-hours).

As a consequence of the investment changes in Norway and Finland, the electricity generation increases for wind in Norway and decreases for coal in Finland. This also results in the CO_2 emission decrease seen in Figure 4.19. A slight CO_2 emission increase is experienced in Denmark, due to a slightly higher level of use of coal and slightly lower level of investments and, thus, use of wind power. This is likely due to the wind power penetration being as large as beneficial with the remaining flexibility in the system. Hence, coal power takes over a small part of the production.

4.9   Discussion and Conclusion

The results are the optimal investments in the situation where all people are rational and acting according to the overall optimum. This is, of course, not always the case, and the modelling cannot capture all individual thinking and acting, but only give an indication to act upon.

From running the model on the Nordic power and road transport system it is obvious that investments in PHEVs are optimal except in scenarios with a low oil price. Furthermore, PHEVs are beneficial for the power system in all the countries no matter the configuration of the power system. They provide flexibility in terms of flexible charging and introduce large savings in the power and transport system. With inclusion of V2G they are an even greater benefit, although the overall costs savings of €18×10^6 are small compared with the savings of €6.2×10^9 from introduction of G2V. The benefits of the flexibility are also reflected in the electricity prices.

Important to notice is the fact that introduction of EDVs is beneficial for both fluctuating production and base load, as seen in the results from Germany. Furthermore, introducing EDVs along with targets for sustainable energy is a good way to ensure more sustainability in the energy system as a whole.

In short, the conclusions are:

- EDVs can help integrate large shares of renewable power sources.
- EDVs can reduce CO_2 emissions for the transport system as well as the power system.

- Overall system costs in the integrated power and transport system can be decreased with introduction of EDVs.
- Power generation on fossil-fuelled power plants benefits from the introduction of EDVs in terms of a more stable power production. This is evident from the results of power generation and electricity prices experienced in Germany.
- With flexibility provided by hydro power, the EDVs are still beneficial.
- The flexibility provided by PHEVs is sufficient for most changes towards renewable energy production in the power system. Thus, changes in investments when introducing 25% BEVs are primarily changes needed to support the increase in electricity demand.

Hence, integration of the power and road transport systems introduces flexibility for the power system as well as incentives for the road transport system to move towards EDVs.

4.10 Summary

In this chapter, an optimization model for road transport has been introduced. This model has been integrated with an existing energy system analysis model, Balmorel, enabling analyses of the configuration of and the investments in the integrated power and road transport system. Integration of the two systems takes into account the benefits of being able to control charging and maybe even discharging of EDVs, both in terms of cost savings and renewable energy integration.

Results show that EDVs can provide flexibility to the power system, resulting in an increase in renewable energy sources in the optimal investments. On the other hand, the transport system also benefits from the integration, in terms of EDVs becoming profitable.

References

Brooks, A. (2002) Vehicle-to-grid demonstration project: grid regulation ancillary service with a battery electric vehicle, *Final report*, AC Propulsion.

Danish Energy Authority (2010) Technology data for energy plants, http://www.ens.dk/Documents/Netboghandel%20-%20publikationer/2010/Technology_data_for_energy_plants.pdf, (accessed January 2013).

Danish Statistics (2009) Familiernes bilrådighed (faktiske tal) efter område og rådighedsmønster.

Danish Statistics (2010) Bestanden af personbiler pr. 1. januar efter tid, drivmiddel og egenvægt.

Delucchi, M., Burke, A., Lipman, T., and Miller, M. (2000) Electric and gasoline vehicle lifecycle cost and energy-use model, *Research Report UCD-ITS-RR-99-04*, Institute of Transportation Studies, University of California, Davis.

Denholm, P. and Short, W. (2006) Evaluation of utility system impacts and benefits of optimally dispatched plug-in hybrid electric vehicles, *Technical Report NREL/TP-620-40293*, National Renewable Energy Laboratory.

DTU Transport (n.d.) Transportvaneundersøgelsen.

Ellehauge, K. and Pedersen, T. (2007) Solar heat storages in district heating networks, http://www.preheat.org/fileadmin/preheat/documents/reports/Solar_heat_storages_in_district_heating_networks.pdf, (accessed January 2013).

European Commission, Directorate-General for Energy and Transport (2010) EU energy in figures 2010 – CO_2 emissions from transport by mode, http://ec.europa.eu/energy/publications/doc/statistics/ext_co2_emissions_from_transport_by_mode.pdf, (accessed January 2013).

Ibáñez, E., McCalley, J., Aliprantis, D., et al. (2008) National energy and transportation systems: interdependencies within a long term planning model, in *2008 IEEE Energy 2030 Conference*, IEEE, Piscataway, NJ, pp. 218–225.

Juul, N. (2011) Modelling and analysis of distributed energy systems with respect to sustainable energy – focus on electric drive vehicles, PhD thesis, DTU Management Engineering, ISBN 978-87-92706-15-7.

Juul, N. and Meibom, P. (2011a) Optimal configuration of an integrated power and transport system. *Energy*, **36**, 3523–3530.

Juul, N. and Meibom, P. (2011b) Road transport and power system scenarios for northern Europe in 2030. *Applied Energy*, **92**, 573–582.

Karlsson, K. and Meibom, P. (2008) Optimal investment paths for future renewable based energy systems – using the optimisation model Balmorel. *International Journal of Hydrogen Energy*, **33**, 1777–1787.

Kempton, W. and Kubo, T. (2000) Electric-drive vehicles for peak power in Japan. *Energy Policy*, **28**, 9–18.

Kempton, W. and Tomic, J. (2005a) Vehicle-to-grid power fundamentals: calculating capacity and net revenue. *Journal of Power Sources*, **144**, 268–279.

Kempton, W. and Tomic, J. (2005b) Vehicle-to-grid power implementation: from stabilizing the grid to supporting large scale renewable energy. *Journal of Power Sources*, **144**, 280–294.

Kiviluoma, J. and Meibom, P. (2010) Influence of wind power, plug-in electric vehicles, and heat storage on power system investments. *The Energy Journal*, **35**, 1244–1255.

Kristoffersen, T., Capion, K., and Meibom, P. (2011) Optimal charging of electric drive vehicles in a market environment. *Applied Energy*, **88**, 1940–1948.

Letendre, S. and Kempton, W. (2002) The V2G concept: a new model for power? *Public Utilities Fortnightly*, (February 15), 16–26.

Lund, H. and Kempton, W. (2008) Integration of renewable energy into the transport and electricity sectors through V2G. *Energy Policy*, **36**, 3578–3587.

McCarthy, R., Yang, C., and Ogden, J. (2008) Impacts of electric-drive vehicles on California's energy system, *Report UCD-ITS-RP-08-24*.

Moura, F. (2006) Driving energy system transformation with "vehicle-to-grid" power, *Interim Report IR-06-025*.

Münster, M. and Meibom, P. (2011) Optimization of use of waste in the future energy system. *Energy*, **36**, 1612–1622.

Nguyen, F. and Stridbaek, U. (2007) *Tackling Investment Challenges in Power Generation: In IEA Countries*, International Energy Agency, Paris, http://www.iea.org/publications/freepublications/publication/tackling_investment.pdf, (accessed January 2013).

Ravn, H. (2010) The Balmorel model structure, version 3.01, http://www.eabalmorel.dk/files/download/TheBalmorelModelStructure-BMS301.pdf, (accessed January 2013).

Ravn, H., Hindsberger, M., Petersen, M., et al. (2001) Balmorel: a model for analyses of the electricity and CHP markets in the Baltic Sea region, http://balmorel.com/Doc/B-MainReport0301.pdf, (accessed January 2013).

Short, W. and Denholm, P. (2006) A preliminary assessment of plug-in hybrid electric vehicles on wind energy markets, *Technical Report NREL/TP-620-39729*.

Tomic, J. and Kempton, W. (2007) Using fleets of electric-drive vehicles for grid support. *Journal of Power Sources*, **168** (2), 459–468.

5

Optimal Charging of Electric Drive Vehicles: A Dynamic Programming Approach

Stefanos Delikaraoglou,[1] Karsten Capion,[1] Nina Juul[2] and Trine Krogh Boomsma[3*]

[1]*Technical University of Denmark, Lyngby, Denmark*
[2]*Department of Management Engineering, Technical University of Denmark, Roskilde, Denmark*
[3*]*Corresponding author, Department of Mathematical Sciences, University of Copenhagen, Copenhagen East, Denmark*

5.1 Introduction

Currently, the transport sector relies on 95% liquid fossil fuels and is responsible for 25% of total greenhouse gas emissions related to energy; see WWF (2008). The potential to reduce this fossil fuel dependency and thereby also emissions, has served as a major motivation for the introduction of electric-drive vehicles (EDVs. For the precise definition, see Section 5.2).

In addition to the above, EDVs may offer flexibility in charging and discharging that is highly valuable in short-term power system operation. Very importantly, the introduction of electric vehicles may help the integration of fluctuating renewable production such as wind power; that is, by charging when wind production is high and discharging when low. This balancing is possible on an hour-by-hour basis (e.g., through the electricity spot market), but may also include the provision of manual and automatic regulation and reserve power on a minute basis.

Despite the potential, the flexibility actually provided depends on the economic incentives for the vehicle operators, where such incentives may be price signals from the electricity

Grid Integration of Electric Vehicles in Open Electricity Markets, First Edition. Edited by Qiuwei Wu.
© 2013 John Wiley & Sons, Ltd. Published 2013 by John Wiley & Sons, Ltd.

market. This serves as our motivation for studying the optimal charging and discharging of EDVs in response to market prices.

The overall aim of this chapter is:

- to suggest an efficient algorithm for the short-term management of an EDV or a fleet of EDVs in a market environment;
- (assuming that electric vehicles operate according to this algorithm) to investigate the flexibility provided to the power system.

Hence, although we take the perspective of the vehicle operators, we also aim to analyze the operation of the electricity system, and in particular the contribution of EDVs to system flexibility.

For vehicle operators participating in the electricity spot market, the problem is to optimally charge and discharge the vehicles, while taking into account the driving demands of the vehicle owners and the variations in electricity spot prices. We consider the cases of a vehicle owner who is a price-taker and of a fleet operator who can influence prices. In both cases, the vehicle operators minimize the costs of charging and discharging by buying and selling electricity in the spot market subject to a number of operational constraints. We focus on short-term management and restrict attention to a weekly time horizon with hourly resolution. With only a weak link between consecutive hours of the week, we show how the problem is amenable to dynamic programming with linear or quadratic costs.

For a single vehicle owner, the change in electricity load due to charging and discharging is sufficiently small not to influence spot prices. An operator of a considerable vehicle fleet, however, may be able to affect prices by changing the load. For simplicity, we assume that prices vary linearly with the electricity load and describe the price–load dependency by linear regression.

In the dynamic programming problem, we discretize the state space. This is computationally feasible for a single vehicle. For a vehicle fleet, however, the problem is prone to suffer from the curse of dimensionality, and to facilitate computations, therefore, we propose an ex ante vehicle aggregation approach. The approach is based on a grouping of the vehicles according to day of departure and intra-day driving patterns. As a result, an aggregate vehicle represents a whole group of vehicles with similar driving patterns.

We illustrate the results of the dynamic programming problem in a Danish case study, using spot prices from the Nordic electricity market and survey data for driving patterns of the vehicle fleet in Western Denmark. Our results show that EDVs have incentive to provide flexibility almost exclusively through charging. Moreover, the vehicles offer significant intra-day flexibility, but only limited day-to-day flexibility.

For other contributions from the literature on EDVs, we refer to Kempton and Tomic (2005a) for general business models and implementation issues regarding the control of electric vehicles, and to Kempton and Tomic (2005b) and Tomic and Kempton (2007) for the provision of grid support, and regulation and reserve power. Closer in spirit to this chapter, Juul and Meibom (2009), Kiviluoma and Meibom (2010), and Shortt and O'Malley (2009) formulate mathematical programming models for analyzing the impact of electric vehicle integration on electricity system operation. However, although similar in their use of mathematical programming models, these references apply social welfare optimization, whereas we take a market-based approach to optimizing private welfare as in Kristoffersen et al. (2011). In all of these references, the mathematical programming problems are solved by standard software tools and no additional efforts are made to improve computational efficiency, whereas this chapter suggests a solution algorithm specially adapted to the control of electric vehicles.

The chapter is structured as follows. Section 5.2 gives a short introduction to EDVs. A discussion of short-term management of such vehicles and the corresponding dynamic programming problems are presented in Sections 5.3–5.5, and we describe electricity price variations and the vehicle aggregation approach in Sections 5.6 and 5.7, respectively. We present the case study in Section 5.8 and report on computational results in Section 5.9. Finally, we suggest improvements and extensions of the dynamic programming problem in Section 5.10.

5.2   Hybrid Electric Vehicles

EDVs comprise all vehicle technologies with an electric drive train. Most importantly, these include battery electric vehicles (BEVs), plug-in hybrid electric vehicles (PHEVs), and fuel-cell vehicles (FCVs). For an extensive overview on EDV technology, see also Smets *et al.* (2007).

In the following, we restrict attention to PHEVs.[A] There are several reasons for this.

Relatively many problems remain to be solved before the commercialization of FCVs can be realized. Furthermore, although BEVs have been on the market for several decades, these vehicles are still not fully competitive. However, several large car manufacturers have announced the introduction of PHEVs in this and the following decade. Examples include the Chevrolet Volt, which was on the market in 2010, as well as the Toyota Prius, launched in 2012, and the Volkswagen Golf, the launch of which is expected in 2015; see Wikipedia (n.d.).

From an electricity system perspective, FCVs are less interesting as such vehicles are still without plug-in capabilities, most likely for economic reasons. In contrast, both BEVs and PHEVs can be plugged in and charged by consuming power from the grid, also referred to as grid-to-vehicle; see Kempton and Tomic (2005a,b).

BEVs rely solely on battery power for propulsion. Even though most trips can be completed without depleting the batteries, substantial charging times render these vehicles less suited for long trips where recharging is necessary. In contrast, PHEVs have internal combustion engines, which allow for longer trips.

We make a few assumptions for the PHEVs considered. Although the internal combustion engines may run on hydrogen, ethanol, or gas, liquid fossil fuels are the most widely used, and we confine ourselves to these. The use of the battery is usually less expensive than running on the engine. The PHEV driving mode, therefore, is often either a charge-depletion mode, in which the battery is used until depleted, or a charge-sustaining mode, in which the engine assists the battery. We assume that no such mode has been predetermined, but allow the optimization to determine the use of the battery versus the engine. Finally, although current development does not allow for vehicle-to-grid (Kempton and Tomic, 2005a,b), we investigate whether there is economic incentive for PHEVs to discharge by supplying power to the grid.

5.3   Optimal Charging on Market Conditions

We consider the short-term management of EDVs in a market environment. From the perspective of vehicle operators participating in the electricity spot market, the problem is to

---

[A] As most of our analysis applies to EDVs in general, we mostly refer to the PHEVs as electric vehicles.

optimally charge and discharge the vehicles, while taking into account the driving demands of the owners and variations in electricity spot prices.

With market participation, charging and discharging of electric vehicles occur through purchases and sales in the spot market. Electric vehicles may have direct access to this market, although market participation may be facilitated by a larger trader (who we refer to as a fleet operator) because of transaction costs and minimum trading sizes. This can be implemented in different ways, as suggested by Kempton and Tomic (2005b) and further discussed in Kempton and Tomic (2005a) and Tomic and Kempton (2007). We consider the cases of a single vehicle owner and that of a fleet operator who buys and sells electricity from and to the spot market on behalf of a number of vehicles owners. We assume that the vehicle owner is a price-taker, but allow the fleet operator to influence market prices.[B]

We make the following assumptions about the flexibility to charge and discharge. Owing to fixed working hours, the vehicle owners are rather inflexible in terms of driving; therefore, we assume that driving demand is fixed exogenously. We further assume that the vehicles are always plugged in when parked, and so charging and discharging is always feasible at such times. Consistent with this, the fleet operator is unable to control the driving patterns of the vehicles, but is able to fully control the charging and discharging when parked.

The problem of obtaining an optimal short-term plan for charging and discharging the vehicles can be formulated as a mathematical program in which the vehicle operator minimizes charging and discharging costs subject to a number of operational constraints. We start by formulating the problem for a single vehicle owner and later extend this formulation for a vehicle fleet.

We may consider a time horizon $[0, T]$ of, for example, a day, a week, or a month and discretize this into, for example, hourly or half-hourly intervals $[t, t + 1]$, $t = 0, \ldots, T - 1$. Given the availability of the data, however, we restrict attention to a weekly time horizon with hourly resolution.

Accordingly, we let the decision variable l_t represent the battery state of charge at time t. We assume that the initial and terminal states l_0 and l_T are known parameters.[C] We likewise let the variables u_t^+ and u_t^- represent the rates of charging and discharging the battery between times t and $t + 1$, and we denote by v_t the engine supply between times t and $t + 1$.

The battery state is constrained by minimum and maximum states of charge given by the parameters l^{min} and l^{max}. In the same fashion, charging and discharging are restricted by maximum rates $u^{+,max}$ and $u^{-,max}$, whereas the engine supply is constrained by the capacity of the engine v^{max}.

To account for charging losses, we let the parameter $0 < \eta < 1$ represent the efficiency of charging. Similarly, we denote by $0 < \rho < 1$ the efficiency of the engine.[D]

Short-term driving demands of the vehicle owners are largely predictable owing to fixed working hours of commuters, and fixed business schedules and routes, which is our reason for assuming a deterministic driving pattern. We denote the driving demand between times t and $t + 1$ by d_t. To capture the corresponding plug-in pattern, we let $\delta_t = 0$ when the vehicle is

---

[B] The price-taker assumption is justified by the results, which show that a small vehicle fleet has little effect on prices.
[C] To avoid undesired end effects, such as the tendency to deplete the battery by the end of the time horizon, we have assumed an exogenously fixed terminal battery state of charge. For example, considering weekly cycles, we may let $l_T = l_0$. An alternative to this would be to estimate the future value of an endogenously determined terminal battery state, using electricity and fuel forward and futures prices.
[D] Note that the formulation also applies for BEVs without an engine by letting the efficiency $\rho = 0$.

parked and plugged in (i.e., for $d_t = 0$) and $\delta_t = 1$ when driving (i.e., for $d_t > 0$). Electricity spot prices are less predictable; therefore, we describe these by a stochastic process $\{p_t\}_{t=0}^{T-1}$ on some probability space. We will assume that this process is a Markovian. In our formulation of the problem, the expected driving demand d_t and the predicted price p_t between times t and $t + 1$ are both known at time t prior to making decisions for the time interval between t and $t + 1$.

For every time interval, the vehicle operator has to make the following decisions: when the vehicle is parked, whether to charge or discharge and at which rate; and when driving, whether to use the engine and likewise how much. The optimal decision is determined by the costs of charging, discharging, and using the engine. We denote by $c(u^+, u^-, v)$ the operating costs in excess of spot market purchase and selling costs. In particular, assuming that these costs consist of battery wear costs and fuel costs, we use the specification $c(u^+, u^-, v) = au^- + bv$, where $a, b > 0$.[E] It should be remarked, though, that since engine supply is usually more expensive than charging, the optimal decision is primarily determined by a trade-off between the current and future spot prices.

The mathematical programming problem for optimal charging is

$$
\min \left\{ \mathbb{E}\left[\sum_{t=0}^{T-1} (p_t(u_t^+ - u_t^-) + c(u_t^+, u_t^-, v_t)) \right] : \right.
$$
$$
l_{t+1} = l_t + (1 - \delta_t)(\eta u_t^+ - u_t^-) + \delta_t(\rho v_t - d_t), l^{\min} \le l_t \le l^{\max}, \tag{5.1}
$$
$$
\left. 0 \le u_t^+ \le u^{+,\max}, 0 \le u_t^- \le u^{-,\max}, 0 \le v_t \le v^{\max}, t = 0, \dots, T - 1 \right\}
$$

where $\mathbb{E}[\cdot]$ denotes the expectation with respect to $p_0, \dots, p_{T-1}$. The constraints include balance constraints for the battery state of charge and capacity restrictions for both the battery state and the rates of charging, discharging, and engine supply.

5.4   Dynamic Programming

The optimal charging problem contains only a weak link between consecutive hours of the week, namely through the balance constraints for the battery state, which makes it amenable to dynamic programming. To rewrite the problem, we let the state variables be the battery state of charge and the electricity price, and we let the controls be the rates of charging, discharging, and using the engine. The dynamic programming recursion is then

$$
\mathcal{J}_t(l_t, p_t) = \min\{p_t(u_t^+ - u_t^-) + c(u_t^+, u_t^-, v_t) + \mathbb{E}_t[\mathcal{J}_{t+1}(l_{t+1}, p_{t+1})] :
$$
$$
l_{t+1} = l_t + (1 - \delta_t)(\eta u_t^+ - u_t^-) + \delta_t(\rho v_t - d_t), l^{\min} \le l_t \le l^{\max}, \tag{5.2}
$$
$$
0 \le u_t^+ \le u^{+,\max}, 0 \le u_t^- \le u^{-,\max}, 0 \le v_t \le v^{\max}\}, \qquad t = 1, \dots, T - 1
$$

where $\mathcal{J}_t$ is the value function of the stage-t problem and $\mathbb{E}_t[\cdot]$ denotes the expectation with respect to p_{t+1} conditional on p_t. The terminal condition is $\mathcal{J}_T(l_T, p_T) = 0$.

[E] Note that this specification of excess costs ensures that it is never optimal to charge and discharge at the same time.

To facilitate computations, we discretize the state space. In particular, we discretize the feasible range of the battery state $[l^{\min}, l^{\max}]$ into intervals $[l^i, l^{i+1}]$, $i = 0, \ldots, I - 1$, with $l^0 = l^{\min}$ and $l^I = l^{\max}$. We solve the stage-t problem by complete enumeration of the solutions, assuming that the states $l_t \equiv l_i$ and $l_{t+1} \equiv l_j$ at times t and $t + 1$ are known. From the balance constraints, we obtain the solution

$$u_t^{+,i,j} = (1 - \delta_t)(l^j - l^i)\eta^{-1} \vee 0, \quad u_t^{-,i,j} = (1 - \delta_t)(l^i - l^j) \vee 0,$$
$$v_t^{i,j} = \delta_t(l^j - l^i + d_t)\rho^{-1}$$

provided this solution complies with the lower and upper bounds on the controls; that is, the capacity restrictions on charging, discharging, and engine supply. We therefore let $\mathcal{U}_t(i) = \{j : u_t^{+,i,j} \leq u^{+,\max}, u_t^{-,i,j} \leq u^{-,\max}, 0 \leq v_t^{i,j} \leq v^{\max}\}$. Note that this solution always generates transitions between the discretized states such that no interpolation between these states is required.[F] The stage-t problem can now be rephrased as

$$\mathcal{J}_t(l^i, p_t) = \min_{j \in \mathcal{U}_t(i)} \{p_t(u_t^{+,i,j} - u_t^{-,i,j}) + c(u_t^{+,i,j}, u_t^{-,i,j}, v^{i,j}) + \mathbb{E}_t[\mathcal{J}_{t+1}(l^j, p_{t+1})]\},$$
$$t = 0, \ldots, T - 1$$

which requires $(I + 1)^2$ evaluations of the stage-t costs, each evaluation involving only the computation of the solution and a feasibility check. It should be remarked that the states and controls are always feasible in the continuous problem, and so the discrete formulation provides an upper bound. For the discretization of electricity prices, see Section 5.6.

5.5   Fleet Operation

We extend the dynamic programming formulation (5.2) for a fleet of vehicles. Accordingly, we denote by $k = 1, \ldots, K$ the vehicles of the fleet and let the K-dimensional vector of decision variables $l_t = (l_{1t}, \ldots, l_{Kt})$ represent the battery states of charge, where l_0 and l_T are known. In a similar fashion, we let the vectors u_t^+ and u_t^- represent the charging and discharging rates to and from the batteries, and we denote by v_t the supply from the engines. Upper and lower bounds on these variables are given by the vectors of parameters $l^{\min}$, $l^{\max}$, $u^{+,\max}$, $u^{-,\max}$, and $v^{\max}$. Charging and engine efficiencies are included through the diagonal $K \times K$-matrices $H = \text{diag}(\eta_1, \ldots, \eta_K)$ and $P = \text{diag}(\rho_1, \ldots, \rho_K)$, whereas driving patterns are given by the vectors $d_t = (d_{1t}, \ldots, d_{Kt})$, and plug-in patterns by the matrices $\Lambda_t = \text{diag}(\delta_{1t}, \ldots, \delta_{Kt})$. As a similar extension of the above, excess operating costs are given by the vector function $c(u^+, u^-, v) = (c_1(u^+, u^-, v), \ldots, c_K(u^+, u^-, v))$, where $c_k(u^+, u^-, v) \equiv c_k(u_k^+, u_k^-, v_k)$; that is, the operating costs of one vehicle are independent of those of other vehicles. In particular, we assume that operating costs are additive. For a fleet, we allow for electricity prices to vary with the total net load of the vehicles. Hence, we specify the price as a function $\hat{p}(p, q)$ of the load q and the price p when ignoring load effects. To compute total costs and load, we introduce the K-vector $e = (1, \ldots, 1)$.

[F] This would not be the case for BEVs without an engine.

Now, the dynamic programming formulation of the vehicle fleet problem is

$$\mathcal{J}_t(l_t, p_t) = \min\{\hat{p}(p_t, q_t) + e^{\mathrm{T}}c(u_t^+, u_t^-, v_t) + \mathbb{E}_t[\mathcal{J}_{t+1}(l_{t+1}, p_{t+1})]:$$
$$q_t = e^{\mathrm{T}}(u_t^+ - u_t^-), l_{t+1} = l_t + (I - \Lambda_t)(Hu_t^+ - u_t^-) + \Lambda_t(Pv_t - d_t), \quad (5.3)$$
$$l^{\min} \le l_t \le l^{\max}, 0 \le u_t^+ \le u^{+,\max}, 0 \le u_t^- \le u^{-,\max}, 0 \le v_t \le v^{\max}\},$$
$$t = 0, \ldots, T - 1$$

where the terminal condition is $\mathcal{J}_T(l_T, p_T) = 0$.

Note that this problem is not separable with respect to vehicles.

5.6 Electricity Prices

When participating in the electricity spot market, charging and discharging should be planned such as to take into account the variations in market prices. In the Danish case study, we consider the electricity spot market Elspot organized by the Nordic power exchange Nord Pool. Elspot is used for trading hourly contracts,[G] and all trades are settled at a market price determined by the balance between demand and supply bids from all market participants. To describe the variations in electricity spot prices, we use historical data for the area price (DKK/MWh) in Western Denmark 2006–2007 provided by Energinet.dk (www.energinet.dk).

5.6.1 A Markov Chain for Electricity Prices

We first describe the stochastic process $\{p_t\}_{t=0}^{T-1}$ of electricity prices. To facilitate the application of dynamic programming, we assume that this process is a Markov chain. For estimation, we compute the mean μ and standard deviation σ of historical prices and discretize the interval $[0, \mu + 2\sigma]$ into intervals (bins) of equal length, $\mathcal{B}_h = [\bar{p}_h, \bar{p}_{h+1}], h = 0, \ldots, H - 1$, with $\bar{p}_0 = 0$ and $\bar{p}_H = \mu + 2\sigma$. We record the frequency with which the historical price makes a transition between any two bins h and k between times t and $t + 1$; that is, $n_{hkt} = \#\{\tau : p_\tau \in \mathcal{B}_h, p_{\tau+1} \in \mathcal{B}_k\}/\#\{\tau : p_\tau \in \mathcal{B}_h\}$,[H] where $h, k = 0, \ldots, H - 1, t = 1, \ldots, T - 1$. The midpoints of the bins $(\bar{p}_{h+1} - \bar{p}_h)/2, h = 0, \ldots, H - 1$, represent the states of our Markov chain, and the frequencies $n_{hkt}, h, k = 0, \ldots, H - 1, t = 1, \ldots, T - 1$ provide us with maximum likelihood estimates for the state transition probabilities.

5.6.2 The Price–Load Dependency

We next describe how electricity prices may be affected by the total net load from a vehicle fleet. For simplicity, we assume that the price is a linearly increasing function of load; that is:

$$\hat{p}(p_t, q_t) = p_t + \gamma q_t$$

[G] Elspot is a day-ahead market for trading contracts with physical delivery the following day. However, in contrast to the day-ahead market clearing, we assume that the market is cleared an hour ahead of operation, or for some other reason the price of the following hour can be predicted, whereas future prices are unknown.

[H] With a slight abuse of notation, p_t now represents the historical price at time t.

where p_t is the (stochastic) price between times t and $t + 1$ when ignoring the effect of the vehicles and γ is a known parameter, representing the sensitivity of the price with respect to the load of the fleet. The parameter γ is estimated by regression. In particular, we regress electricity spot prices on total electricity load of the power system, using hourly consumption data provided by Energinet.dk. Ordinary least-squares estimation gives us an estimate of 0.18 (DKK/MWh)/MWh with a t-statistic of -43.69 and a p-value of 0.00;[1] see also Kristoffersen *et al.* (2011).

In the current power system, electricity is largely nonstorable, and supply and demand must therefore balance at all times. This may occasionally result in price jumps in situations of grid congestion; that is, for very high (low) loads, import (export) transmission restrictions may produce significant price increases (decreases). The introduction of electric vehicles and the accompanying provision of storage may facilitate a partial decoupling of supply and demand. However, if price jumps still occur, the price–load dependency may be more appropriately described by a nonlinear function.

5.7   Driving Patterns

To successfully plan charging and discharging, the construction of representative driving patterns is crucial. We assume that the driving patterns of electric vehicles are similar in nature to historical driving patterns. In the Danish case study, therefore, we use survey data on the vehicle fleet in Western Denmark 2006–2007 provided by the Department of Transport, Technical University of Denmark (DTU, 2009). The data cover passenger transport activities and include a large number of variables, such as time (hours), location, and distance (kilometers). However, we do not consider locational differences, and the data do not allow us to distinguish between private and business driving. To construct the driving patterns of PHEVs in particular, we further assume that these vehicles are represented by those in the survey driving more than 150 km in a day.

5.7.1   Vehicle Aggregation

In the dynamic programming problem, we discretize the state space. This is computationally feasible for a single vehicle. For a vehicle fleet, however, the problem is prone to suffer from the curse of dimensionality; therefore, to facilitate computations, we propose an ex ante vehicle aggregation approach.

We start by constructing daily driving patterns. For each vehicle (we assume a single passenger per vehicle and vice versa) and each day covered by the survey, we record the day (Monday–Sunday) and the hour (1–24) of departure for all trips. For each trip, we record the driving distance in the hour of departure and, where relevant, we calculate the driving distance in subsequent hours, assuming an average speed of 70 km/h. A daily driving pattern is stored as a 24-dimensional vector of distances and has an associated weekday. This results in a total of 561 driving patterns.

We then group the daily driving patterns according to day of departure. The driving patterns reveal commuter behavior on weekdays with departure peaks between 6 and 9 p.m. and return peaks between 3 and 6 p.m., whereas weekend days and holidays show no peaks.

---

[1] This estimate is in fact obtained in a multivariate regression of prices on electricity demand, wind and hydropower production, and coal, gas, and emission prices.

By distinguishing only between weekday and weekends, this produces 415 weekday-driving patterns and 146 weekend patterns (the reason that 415/561 is not exactly equal to 5/7 is the assumption that PHEVs are represented by those vehicles in the survey driving more than 150 km in a day).

We continue by clustering the intra-day driving patterns within each group, applying a variation of the k-means algorithm; see McQueen (1967). This algorithm obtains k clusters by grouping the vectors according to similarity, which we measure as being sufficiently close in terms of some distance. The algorithm proceeds as follows. Select k vectors as initial centers of the clusters. For each of the remaining vectors, determine the center closest in distance and assign the vector to the corresponding cluster. Update the centers and repeat the assignment to clusters. We ran the clustering algorithm with different centers and distance measures. When we choose the center as the mean vector of each cluster, driving takes place in most hours of a day, whereas in the historical data it takes place only in some hours of the day. To overcome this problem, we replace the center by the vector closest to the remaining vectors of a cluster, and confirm that driving takes place only in some hours of the day. When we choose the Euclidian distance, the algorithm produces unrealistic driving patterns by not taking into account dependencies between subsequent hours of driving; therefore, we define a new distance measure that is small when the driving distance in subsequent hours is almost the same. The clustering algorithm was implemented in C++ and run with a total of 561 original and 20 representative driving patterns, that is, 10 weekday and 10 weekend patterns. The results show that, in spite of a rather crude clustering, the driving patterns are reasonably representative. For further details on the vehicle aggregation, see Kristoffersen *et al.* (2011).

To reduce the dimension of the state space in the dynamic programming problem, we now consider aggregate vehicles $k = 1, \ldots, K$ in (5.2). We denote for each aggregated vehicle k the number of vehicles it represents by n_k, and replace the vector e by $n = (n_1, \ldots, n_K)$.

5.8   A Danish Case Study

To illustrate the short-term management of EDVs, we use a Danish case study. The driving patterns and electricity prices were obtained in Sections 5.6 and 5.7, whereas the remaining assumptions and data are provided below.

Although most likely EDVs will be marketed with several battery sizes, for illustration purposes we confine ourselves to one size. The PHEVs are assumed a 65 km all-electric driving range, which complies with most demonstration types; see Bradley and Frank (2009). Driving efficiencies vary substantially in the literature. However, as in Suppes (2006), we use an efficiency of 6 km/kWh, which corresponds to a PHEV battery size of 10.8 kWh.

For further capacity data, we use the AC-150 Gen-2 vehicle (AC Propulsion, n.d.) that allows for 20 kW charging and discharging from and to the battery. Furthermore, we assume the vehicle owners have access to home and workplace charging through the current grid; in the Danish case this is a three-phase 16 A 230 V system, such that charging and discharging maximum capacities from and to the grid are in fact 11.1 kW, which is in the range of the 10–15 kW found in Kempton and Tomic (2005a). Minimum capacities are set to 20% of maximum capacities.

To estimate fuel costs, we assume 1 L of diesel amounts to 10 DKK, including taxes, and has an energy content of 10 kWh such that the resulting energy costs are 1000 DKK/MWh. The costs of vehicle-to-grid battery wear are set to 394 DKK/MWh, which complies with Kempton and Tomic (2005a).

The charging efficiency is assumed to 90% (Kempton and Tomic, 2005a), whereas the efficiency of the engines is set to the 2015 peak efficiency of 39% or 23.4 km/L; see Smets *et al.* (2007).

5.9 Optimal Charging Patterns

We present the results of the dynamic programming algorithm for a single vehicle and a vehicle fleet. Assuming the vehicles operate as suggested by this algorithm, we use these results to investigate the flexibility provided to the power system.

The dynamic programming algorithm is run on an Intel Core2Duo CPU @ 2.4GHz; RAM: 4 GB.

5.9.1 Single Vehicle Operation

For ease of exposition, we start by analyzing the results from the deterministic formulation of the single vehicle problem. We select a vehicle and two representative weeks, one of them with high prices (average 479 DKK/MWh) and the other with typical prices (average 184 DKK/MWh). In the high-price week, the results show total charging of 179 kWh purchased at an average spot price of 382 DKK/MWh and total discharging of 35 kWh sold at an average spot price of 1260 DKK/MWh. With almost 20% of charging used for discharging, this indicates a potential for using the electric vehicles as storage. The results are very different in the typical week, where total charging is 141 kWh purchased at an average spot price of 176 DKK/MWh and there is no discharging. Clearly, this points towards a very limited storage potential. The different discharge rates in the two weeks can be explained by relatively high battery wear costs, which makes the use of current charging for future discharging unprofitable unless at very high prices. It is worth noting, however, that when the electric vehicles do not provide flexibility to the grid as storage, they may provide flexibility through charging.

The charging patterns can be studied in Figure 5.1 for the high-price week and Figure 5.2 for the typical week. As can be observed from these figures, charging mainly occurs on weekdays. In the high-price week and the typical week, weekday charging accounts for 87% and 83% respectively (in both cases $>5/7 = 71\%$) of total charging and is due to the nature of most working schedules (which results in two longer trips per day, each completed within 1 or 2 h), and longer weekend trips (each completed within 3 h) which cannot be made only on the battery. In spite of the differences between weekdays and weekends, the charging patterns are relatively table from one weekday to another with percentage standard errors of 14% and 9% in the high-price week and the typical week, respectively. For example, the rate of charging remains rather unaffected by higher prices on Friday in both weeks. This can be explained by the fixed driving patterns combined with the limited battery capacity, and implies that the vehicle does not provide day-to-day flexibility to the grid. However, we further observe from the figures that charging mainly occurs during nighttime. Nighttime charging[J] makes up 64% and 92% of total charging in the high-price week and the typical week, respectively, and can be explained partly by the working schedules (or, equivalently, by the availability of the vehicles) and partly by lower prices. By charging mostly at night when the overall demand is low and less frequently during the day when demand is high, the electric vehicle therefore

---

[J] We define the night to be 18:00–06:00.

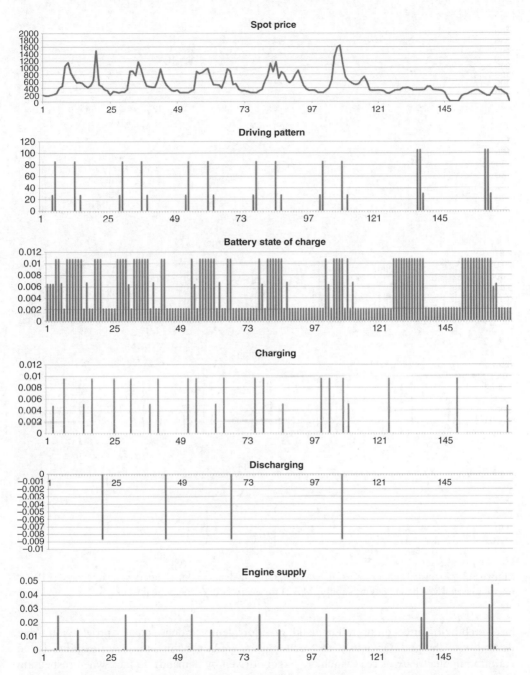

Figure 5.1 Electricity spot prices (DKK/MWh), driving demand (km), battery state of charge (MWh), rate of charge (MWh), rate of discharge (MWh) and engine supply (MWh) for every hour of the high-price week.

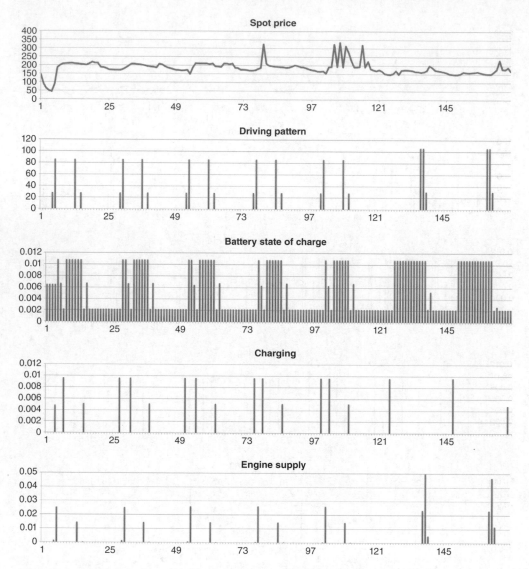

Figure 5.2 Electricity spot prices (DKK/MWh), driving demand (km), battery state of charge (MWh), rate of charge (MWh) and engine supply (MWh) for every hour of the typical week.

has a stabilizing effect on overall demand; or, equivalently, it does actually provide intra-day flexibility to the grid. Obviously, the economic incentive for this intra-day flexibility is the variation in electricity prices (which are largely driven by demand). In fact, when respecting the driving demand and the capacity figures of the vehicle, charging is always determined by a trade-off between current and future prices.

In spite of higher prices, daytime charging does occur, particularly in the high-price week to facilitate future discharging at very high prices, but also in both weeks when driving demand would otherwise exceed the battery capacity, which shows that charging is less expensive than using the engine. This is due to the differences between electricity spot prices and fuel costs (1000 DKK/MWh) and between the charging efficiency (90%) and the efficiency of the engine (39%). The costs of using the engine for driving are $1000/0.39$ DKK/MWh = 2564 DKK/MWh, whereas electricity prices have to be at least 2564×0.9 DKK/MWh = 2308 DKK/MWh for charging to be this expensive. The engine is used only when the driving demand would otherwise not be met, as for example during the longer trips in the weekend.

The results from the stochastic formulation of the single vehicle problem are not as easy to display as those from the deterministic formulation. Only the first-stage solution can be implemented, whereas the solutions of the remaining stages depend on the development of future electricity prices and, thus, give rise to a large number of possible solution paths. We run the dynamic programming algorithm with 10 price states in each stage, which produces 10^{168} such solution paths. By inspection of the average solution path in the first stages, however, we find that, as for the deterministic problem, charging does not necessarily occur immediately prior to driving, but rather when most profitable in terms of the electricity prices. Hence, consistent with deterministic problem, charging is determined by a trade-off between current and *expected* future prices.

5.9.2 Vehicle Fleet Operation

We next consider the results from vehicle fleet operation. As above, for simplicity of the analysis, we confine ourselves to the deterministic formulation of the problem. Moreover, we restrict attention to the typical week.

We initially run the dynamic programming algorithm with three aggregate vehicles obtained from the aggregation approach of Section 6.7, and representing 31, 51, and 63 vehicles. We use 10 battery states for each vehicle, which produces a total of 10^3 battery states in each stage of the problem. With a small fleet of 145 vehicles, however, total load is sufficiently small not to significantly affect electricity prices (the maximum change in the price is 0.16 DKK/MWh). For this reason, the conclusions above remain valid, as can be observed from Figure 5.3.

To show the results for a larger vehicle fleet that is able to influence electricity prices, we continue by scaling the fleet such that the three aggregate vehicles represent 15 500, 25 500, and 31 500 vehicles. This makes up a fleet of 72 500 vehicles, and the load has a significant effect on prices (the maximum change in the price is 29 DKK/MWh).

When comparing the results without and with the price effect,[K] we observe the following. The amount of weekday charging remains almost unaffected (increases from 66% without to 67% with a price effect), which is consistent with the limited day-to-day flexibility provided by the vehicles. However, the intra-day flexibility allows nighttime charging to decrease from 70% to 65%, such that nighttime prices do not increase more than necessary. This is also supported by studying Figure 5.4, which shows that without the price effect it is optimal to fully charge the vehicle at the lowest possible price (given the operational constraints). In contrast, with the price effect, prices increase with the rate of charging or, equivalently, electricity purchase costs are marginally increasing; hence, it is better to partially charge in a number of hours with a low price. As a result, the stabilizing effect of charging on overall

---

[K] Note that we can do this by comparing the results from the smaller and larger vehicle fleet. Without a price effect, which is almost the case for the smaller fleet, the objective function is simply scaled by the size of the fleet, and so the solutions are unaffected by the fleet size.

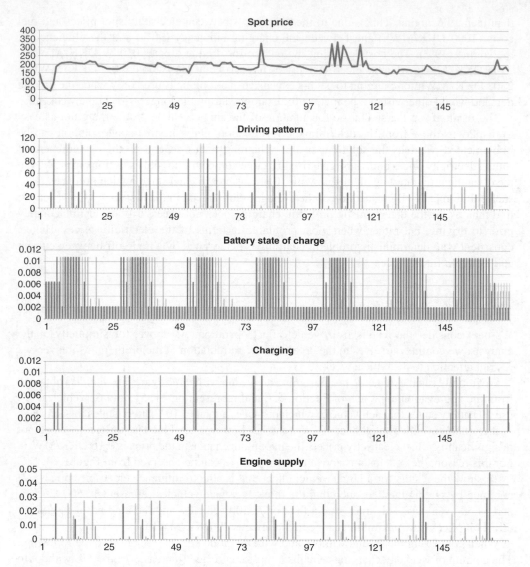

Figure 5.3 Electricity spot prices (DKK/MWh), driving demand (km), battery state of charge (MWh), rate of charge (MWh) and engine supply (MWh) for every hour of the typical week and a fleet of 145 vehicles.

demand may be partially offset by the price effect. In spite of this, there is a small stabilizing effect on electricity prices (the standard deviation decreases from 20% without to 19% with a price effect).

Finally, to investigate the effect of the state space aggregation, we run the dynamic programming algorithm for an increasing number of vehicles. The computation times are 22 s for two vehicles (100 battery states in each stage), 1456 s for three vehicles (1000 battery states), and 102 848 s for four vehicles (10 000 battery states), which proves the aggregation approach to be highly valuable.

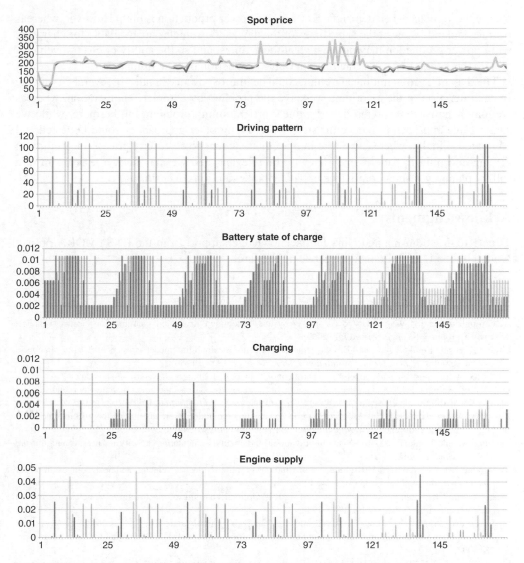

Figure 5.4   Electricity spot prices (DKK/MWh) with (darker) and without (lighter) load-effects, driving demand (km), battery state of charge (MWh), rate of charge (MWh) and engine supply (MWh) for every hour of the typical week and a fleet of 72,500 vehicles.

5.10   Discussion and Conclusion

In this chapter, we suggested an efficient algorithm for the short-term management of EDVs in a market environment and investigated the resulting flexibility provided by such vehicles to the power system. We found that although optimal management of the vehicles does not allow for storage and day-to-day flexibility, the market provides incentive for intra-day flexibility.

Note that electricity prices in fact reflect the variations in net demand; that is, demand less any fluctuating renewable production, such as wind power production. Thus, it is optimal to charge the vehicles when demand is low or wind power production is high. However, whereas overall demand is low at the same times of the day as the vehicles are available for charging, wind power production may not be high at these times, and so the ability of electric vehicles to help the integration of fluctuating renewable production may be limited.

The dynamic programming algorithm could be improved and extended in several directions. In each iteration of the algorithm, we solve the problem by complete enumeration, whereas we may benefit from solving this as a linear programming problem. Furthermore, while we suggest one approach to reduce the state space, many other approaches could be tested and compared. Finally, whereas we take into account stochastic electricity prices, uncertainty in driving and plug-in patterns may likewise be relevant, and we may solve the problem in a rolling planning fashion to produce implementable solutions over time.

Acknowledgments

Trine Krogh Boomsma and Nina Juul acknowledge support from the ENSYMORA project funded by the Danish Council for Strategic Research.

References

AC Propulsion (n.d.) AC-150 Gen-2 EV power system: integrated drive and charging for electric vehicles, www.acpropulsion.com (accessed July 2009).

Bradley, T.H. and Frank, A.A. (2009) Design, demonstration and sustainability impact assessments for plug-in hybrid electric vehicles. *Renewable and Sustainable Energy Reviews*, **13** (1), 115–128.

DTU (2009) Transportvaneundersøgelsen, Department of Transport, Technical University of Denmark.

Juul, N. and Meibom, P. (2009) Optimal configuration of future energy systems including road transport and vehicle-to-grid capabilities, in *2009 European Wind Energy Conference and Exhibition*, EWEC, pp. 168–174.

Kempton, W. and Tomic, J. (2005a) Vehicle-to-grid power implementation: from stabilizing the grid to supporting large-scale renewable energy. *Journal of Power Sources*, **144** (1), 280–294.

Kempton, W. and Tomic, J. (2005b) Vehicle-to-grid power fundamentals: calculating capacity and net revenue. *Journal of Power Sources*, **144** (1), 268–279.

Kiviluoma, J. and Meibom, P. (2010) Influence of wind power, plug-in electric vehicles, and heat storages on power system investments. *Energy*, **35** (3), 1244–1255.

Kristoffersen, T.K., Capion, K. and Meibom, M. (2011) Optimal charging of electric drive vehicles in a market environment. *Applied Energy*, **88**, 1940–1948.

McQueen, J. (1967) Some methods for classifications and analysis of multivariate observations, in *Proceedings of the Fifth Berkeley Symposium on Mathematical Statistics and Probability*, vol. 1, University of California Press, pp. 281–297.

Shortt, W. and O'Malley, M. (2009) Impact of optimal charging of electric vehicles on future generation portfolios, in *2009 IEEE PES/IAS Conference on Sustainable Alternative Energy (SAE)*, IEEE, pp. 1–6, doi: 10.1109/SAE.2009.5534861.

Smets, S., Badin, F. and Brouwer, A. *et al.* (2007) Status overview of hybrid and electric vehicle technology, Technical report, International Energy Agency.

Suppes, G.J. (2006) Roles of plug-in hybrid electric vehicles in the transition to the hydrogen economy. *International Journal of Hydrogen Energy*, **31** (3), 353–360.

Tomic, J. and Kempton, W. (2007) Using fleets of electric-drive vehicles for grid support. *Journal of Power Sources*, **168** (2), 459–468.

Wikipedia (n.d.) List of modern production plug-in electric vehicles, http://en.wikipedia.org/wiki/List_of_modern_production_plug-in_electric_vehicles (accessed January 2013).

WWF (2008) Plugged in: the end of the oil age. Summary report. WWF, Brussels.

6

EV Portfolio Management

Lars Henrik Hansen,[1] Jakob Munch Jensen[1] and Andreas Bjerre[2]
[1]DONG Energy, Power Concept Optimization, Gentofte, Denmark
[2]DONG Energy Sales and Distribution, Virum, Denmark

6.1 Introduction

Electric vehicles (EVs) are considered as distributed energy storage devices which can be used to balance the power fluctuations from renewable energy resources such as wind power, solar power, and so on. In the Danish context, the renewable energy resources is mainly wind power.

A lot of ideas on how to integrate EVs in the electrical power system have been presented in the past. This chapter will give a brief description of how EVs are integrated into the Nordic electricity system today. Further, it will address some of the main issues that a large-scale mobilization of EVs will cause, if the current method is kept. Afterwards, three possible solutions to these challenges will be presented and compared with one another. In the following, the current and alternative methods of integrating EVs are all denoted as changing strategies.

To ensure that the boundary conditions for the analysis are clear, a brief introduction to the Nordic electrical power system will be given, along with a very brief description of the individual actors and their roles.

Since EVs are seen as distributed energy storage devices, it is important to assess their potential flexibility. A prerequisite of all analysis in this chapter is that the primary function of EVs, which is providing their owners with means of transportation, must be honoured at all time. The flexibility of the EVs is heavily influence by the adopted charging strategy, which will be discussed in further detail later in this chapter.

In order to better understand the capabilities of EV portfolio management, a case study is presented. The case study will analyse five different charging methods combined with the significance of the three scenarios: one- or three-phase charging, as well as the penetration level of EVs into the general car population. The case studies are examined using the grid of the Danish island Bornholm as underlying infrastructure.

Grid Integration of Electric Vehicles in Open Electricity Markets, First Edition. Edited by Qiuwei Wu.
© 2013 John Wiley & Sons, Ltd. Published 2013 by John Wiley & Sons, Ltd.

6.2   EV Fleet Modelling and Charging Strategies

6.2.1   System Set-up

Since the Edison project has picked Bornholm as a case study, the power system in focus is constituted by the Nordic power system set-up. In order to clarify which actors exist, what their interests are and how they influence the integration of EVs in this framework, a short description of these actors is presented.

The role of the Transmission system operator (TSO) is to:

– maintain overall system balance;
– ensure fully functioning interconnections to neighbouring countries;
– ensure cost-efficient and reliable operation of the electrical transmission system;
– maintain system frequency stability through the existing power reserves market;
– avoid transmission capacity restrictions due to increase in peak load;
   o increase the amount of demand response power in the system;
   o know the capacity of available ancillary power in the system;
– effectively integrate the increasing amount of wind power into the system;
   o maintain short- and long-term system balance;
– maintain a stable and functioning market for electricity and ancillary services.

The role of the Distribution system operator (DSO) is to:

– distribute and ensure stable supply of electricity to all end costumers within their geographical area;
– perform load shaping and thereby postpon/avoid grid reinforcements and avoid bottlenecks;
– know when and where EVs (or any other load) could cause congestion;
– have a possibility to influence the EV (or any other flexible consumption) induced load in
   the grid;
– be responsibile for metering consumption (and production) on all customers within the area
   of the DSO.

   The DSO and TSO are monopolies, which are governed by the energy authorities. A sketch of the system set-up in the Nordic countries is presented in Figure 6.1. As can be observed, a liberalized market also exists with the following actors:

The role of the Balance Responsible Party (BRP) is:

– In the liberalized energy market, all trading with electricity, be it consumption or production,
   is handled via a BRP. The role of the BRP is mainly to be an insurance agent, so to say.
   They vouch for the individual actors in the electricity market and are obliged to cover
   the actors' deficits, if an actor is unable to pay. In other words, the BRP has the financial
   responsibility for the production and the consumption of its associated actors, which are the
   power producers and the wholesalers.

The role of the Power producers (generation side) is to:

– generate power for the energy exchange market, NordPool. Any production must be provided
   through a BRP, and a power producer is typically also a BRP.
– provide services to the power markets driven by the TSO.

The role of the Wholesalers and retailers (consumption side) is:

– Wholesalers buy energy on the energy exchange market, NordPool, on behalf of their associated retailers. A wholesaler is typically also a BRP.
– Retailers have a contractual relation with the wholesalers. Retailers (or the wholesalers) estimate the expected daily consumption.
– Retailers offer energy to the end customers through contracts, who have the freedom to choose their favourite energy supplier between all the retailers.

The role of the NordPool (energy exchange market) is to:

– serve as energy exchange market for all BRPs in the Nordic system – both generation and consumption.

It should be noted that, since the TSO in Denmark does not possess any production facilities, the TSO has to acquire services for system stability, and so on, through dedicated markets. Today, these markets are served predominantly by the production side, since the consumption side does not really have access to any flexible consumption. However, with the introduction of flexible consumers (e.g. like EVs, heat pumps) this will introduce new possibilities, new actors and maybe new challenges in the power system depicted in Figure 6.1.

Within an unbundled market *set-up* like the Nordic system there are two predominant approaches to handling and mobilizing these new flexible consumers:

• acting on a price signal provided by a BRP (indirect control);
• aggregated and served by a fleet operator (direct control).

Further descriptions and considerations on the above subjects can be found in Hay *et al.* [1].

A correlation study of the spot prices in the Nordic electricity market will reveal that the spot prices are typically low when there is a lot of wind power production and vice versa.

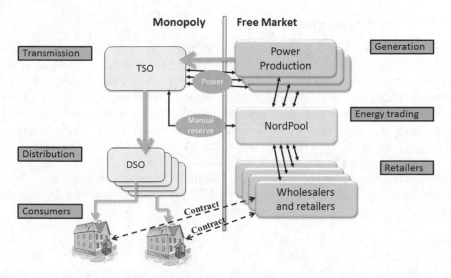

Figure 6.1 The Nordic system set-up with actors: TSO, DSO, consumers, power producing units, the energy exchange market in NordPool and the wholesalers and retailers.

Therefore, a simulation scenario of spot-price-based EV charging has been developed to have the EV charge at the time periods where the spot prices are low. The concept can utilize wind power for EV charging as much as possible and minimize the EV charging cost at the same time. In order to investigate the benefits of the spot-price-based EV charging scenario, three more charging scenarios have been studied as well. These are a dumb charging, timer-based charging and a fleet-operator-based charging scenario.

6.2.2   Battery Modelling

A detailed battery model has been developed describing the charging power as function of time, initial state of charge (SOC) and maximum possible charging current. The solution of the model is analytical and the computations are thus very fast. In the present analysis, it turned out that the bottleneck for the individual EV is always the domestic electrical installation (one-phase 230 V or three-phase 400 V). Furthermore, the batteries are only charged in the range from $SOC = 0.2$ to $SOC = 0.85$. Therefore, the maximum charging power is not limited by the battery, but instead by the electrical installation, and the battery is never charged in the highly nonlinear region at the end of the charging period (above the 'intermediate point' at an SOC close to one). The reason for limiting the charging range to $0.2 \leq SOC \leq 0.85$ is due to the increase in battery life by avoiding deep charge–discharge cycles. It turned out to be sufficient in the present analysis not to include the detailed model and just use a constant maximum charging power. The detailed battery model is included in appendix B of Wu *et al.* [2], as there is a general need for battery models with analytical solutions for fast, detailed computations.

In Figure 6.2, on the left-hand side of the intermediate point, the battery is charged at constant current, whereas on the right-hand side the battery is charged at constant voltage. As the batteries in the present analysis are used in the range from $SOC = 0.2$ to 0.85, the maximum charging power may be assumed to be constant.

6.2.3   Charging Strategies

The EVs are said to be flexible loads. To illustrate the term 'flexibility' in this context, consider an EV with a battery storage capacity of 23 kWh coming home in the afternoon at 17:00 and leaving the next morning at 05:00. It would take 2 h to fully charge the EV using a three-phase 16 A connection (based on 11 kW, which is typical for a Danish household) if the EV returns home nearly empty. This implies that the empty EV could be started to charge any time from 17:00 to 03:00 or even closer to 05:00 depending on the state of charge of the battery. Thus, when and how much to charge is the essence of the EV flexibility – while the method of how to use this flexibility can be referred to as the charging strategy.

The subject of this section is distributed grid charging strategies, which can fit into the existing market structure as described in Section 6.2.1. These charging scenarios are exemplified through the following four charging strategies:

- *Dumb charging (i.e. plug and charge).*   Charging starts immediately after grid connection. Charging continuous until full charge.
- *Timer-based charging.*   Pretty similar to dumb charging, except the start of the charging is programmed through a timer to start outside peak hours.

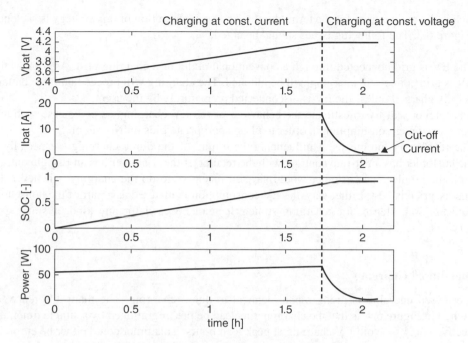

Figure 6.2 Sketch of a detailed charging process.

- *Price-signal-based charging.* Based on an online price signal, the local EV charger can decide to charge (low price) or postpone charging (high price) during grid connection.
- *Fleet-operator-based charging.* Based on service contracts, the fleet operator takes responsibility for the charging of EVs during their grid connection. The fleet operator can then optimize this charging to a wide variety of parameters, such as price or renewable power content.

These four charging strategies are most relevant when seen from a distributed grid load perspective. Emerging alternative charging solutions are exemplified by battery swap stations and fast charge stations, which by their nature will be concentrated in dedicated stations with sufficient grid capacity. From this perspective, they will not cause a threat for the grid stability and, therefore, they are not considered in this analysis.

Further reading on the charging, markets and actors can be found in Ref. [3].

In the following subsections, each of the four charging strategies will briefly be introduced. For simplicity, and without loss of generality, the two actors 'wholesalers' and 'retailer' have been put together in the 'retailer'.

Dumb Charging

The 'plug and charge' strategy will be referred to as dumb charging, as this strategy behaves as a conventional grid load. There is no information exchange with the grid connection point, which is the main reason for the term 'dumb'. The grid integration of this strategy is sketched in Figure 6.3. The following issues should be noted:

- The EV is grid connected through a conventional outlet or a charging post. As soon as the EV is grid connected, charging will be initiated. The charging will proceed until the battery is fully charged unless the EV is disconnected prematurely by the user.
- The retailer will try to estimate the behaviour of the EV charging, as is the case with any other customer consumption in order to place appropriate bids on NordPool.
- The retailer has no means of influencing the timing of the dumb charging – especially if the customer has a flat-rate contract with the retailer. If the customer has an Elspot contract with the retailer, the customer would have an incentive to do the charging, whenever the energy price is best. Either way, the EV consumption is metered as a part of the household consumption. Hence, this charging strategy cannot be used for any kind of smart grid services.

Timer-Based Charging

Timer-based charging is pretty much identical to dumb charging. The main difference, as sketched in Figure 6.4, is that the charging time can be preprogrammed by a simple timer, the objective being to avoid EV charging at peak load hours. This timer could either be enforced using legislation (time tariffs) or could act as the most basic possibility for the user to charge at the cheapest hours of the night. The following issues should be noted:

The EV is grid connected through a conventional outlet or charging post. The EV can be connected, but charging will not take place until grid connection is established by the timer,

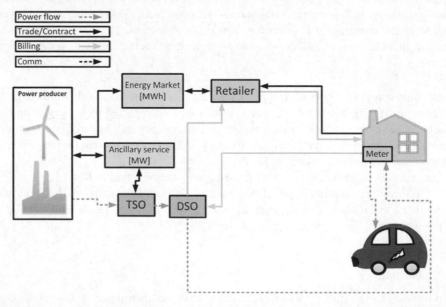

Figure 6.3   Sketch of the dumb charging scenario. House and EV typically have the same owner.

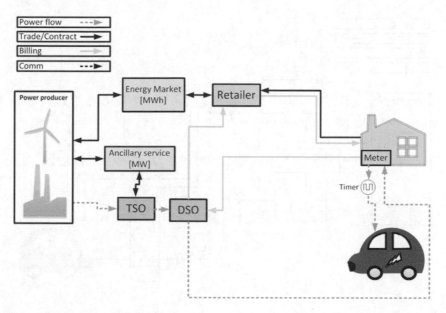

Figure 6.4   Sketch of the timer-based charging scenario. A simple timer is used to preprogram possible charging periods.

and then charging will be initiated. The charging will proceed until the battery is fully charged unless the EV is disconnected prematurely by the timer or user.

- The retailer will try to estimate the behaviour of the timer-based EV charging, as is the case with any other customer consumption. If the retailer were to know the time setting of the timer, they could get a slightly better basis for the estimation than in the dumb charging scenario. However, the retailer can only guess.
- The customer will most likely have a spot price contract with the retailer. The customer would have an incentive to do the charging whenever the energy price is best. If, for example, this happens to take place during the night, the timer can be preprogrammed for the best hours, when the customer is sleeping.

Price-Signal-Based Charging

The price-signal-based charging strategy is sketched in Figure 6.5 using a public charging post as grid connection point. A charging post for the private domain (low-cost version of public version) with similar functionality is assumed to be available for home installation. This strategy involves rather simple information exchange with the charging post. The following issues should be noted:

- The EV is connected to the grid through a public charging post with a dedicated standardized charging cable. The EV will perform a handshake with the charging post to ensure safety. Then the EV will be presented to an online price signal from the responsible retailer and

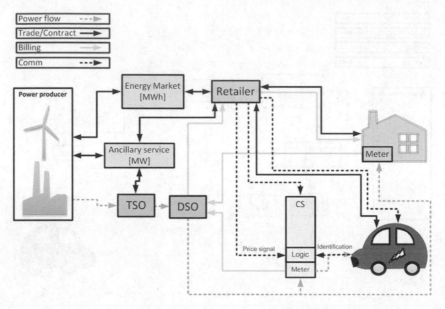

Figure 6.5 Sketch of the price-signal-based charging scenario.

the local EV controller will decide whether to charge or not. The charging will proceed as long as the price is attractive and do so until the battery is fully charged unless the EV is disconnected by the user.

- Depending on the business case set-up of the retailer, the EV user can either pay in advance or after the charging. Either way, the charging process is monitored by a remotely read interval meter.
- The retailers of the public charging posts will try to estimate the behaviour (the load) of all their attached charging posts. Likewise, the retailers of the private home charging posts will also try to estimate the behaviour of their attached charging posts. The purpose of the estimation is to be able to buy an adequate amount of hourly consumption on the day-ahead energy market.
- The retailers have their online price signals as a handle to influence the charging taking place on their charging posts. If the price is increased, the consumption will decrease, and vice versa. This ability can be used by the retailers to participate in the ancillary service markets.
- Due to flexibility of the potential charging response relative to a given price signal, this charging strategy can be used for some kind of future smart grid services. One example of this could be services to the DSO; for example, congestion management. This would imply, however, that the retailer has to be able to distribute localized price signals according to local grid topology.
- In control theory terms, price-signal-based charging performs open-loop control.

As mentioned above, it should be noted that for simplicity the retailer box in Figure 6.5 consists of both retailer and wholesaler. In reality, it will be either the retailer or the wholesaler – depending on the skills of the retailer – who provides the price signal. The price-signal strategy has been omitted in the simulations in this study.

More details on price-signal-based charging can be found in Bang *et al.* [4].

Fleet Operator Control

The fleet-operator-based charging strategy is sketched in Figure 6.6 using a public charging post as a grid connection point. A charging post for the private domain (low-cost version of the public version) with similar functionality is assumed to be available for home installation as well. The fleet-operator-based charging strategy involves information exchange with the charging post as well as information exchange with the fleet operator. The following issues should be noted:

- The EV is grid connected through a charging post with a dedicated standardized charging cable. The EV will perform a handshake with the charging post to ensure safety. Then the EV will negotiate a charging schedule with the fleet operator and start charging according to this schedule.
- Depending on the business case set-up of the fleet operator, the charging schedule could, for example, reflect the current state of charge of the battery and the service level according to the contract between the EV owner and the fleet operator. Either way, the charging process is monitored by a remotely read interval meter.
- The fleet operator has a contract with a retailer. Since the fleet operator has a tight relation to the EV customer, the fleet operator has an excellent opportunity to forecast the expected day-ahead consumption. On an aggregated level this information will be provided to the

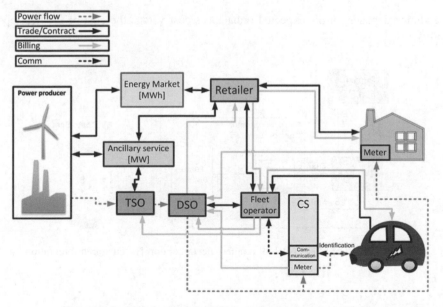

Figure 6.6 Sketch of the fleet-operator-based charging scenario.

retailer. Based on expectations to, for example, available renewable energy resources and other relevant market information, the retailer can to some degree optimize the forecasted consumption to fit the bids forwarded to the Elspot market; that is, move energy consumption to hours with low prices or alternatively hours with a high penetration of renewable energy. When the Elspot[1] is settled, the retailer can send a consumption plan and energy prices to the fleet operator. This scenario is sketched in Figure 6.7.

- The fleet operator of the public charging posts will try to estimate the behaviour (the load) of all their attached charging posts. The strategy is pretty much equivalent to the concept described above – with the minor difference that the public charging posts must be able to serve EV customers from other fleet operators. Here, roaming must be implemented. However, estimating the behaviour of roaming customers will probably be more difficult – less precise.
- The fleet operators have the charging schedules as a handle to influence the charging taking place on their charging posts. If an EV customer behaves different than expected (e.g. wants to get immediate charging) the fleet operator can send a new charging schedule to this customer. If a lot of EV customers behave different than expected, then these customers would have to get new schedules, and it might be necessary to send new schedules to other customers in order to counteract imbalances in relation to the consumption plan provided by the retailer. The part of the fleet operator which is responsible for the dynamic scheduling is called the 'power dispatch' controller in Figure 6.7.
- The retailer has the usual noncontrollable customers and now also a fleet operator with flexible and controllable EV customers. If the retailer has made a poor estimate on the expected consumption, they may either:
  – take the imbalances cost as usual;
  – ask the fleet operator to adjust consumption.

The choice depends on the expected imbalances cost versus the cost of the fleet operator flexibility.

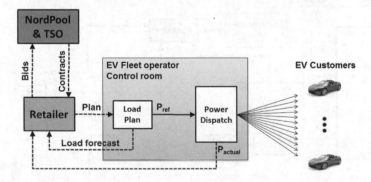

Figure 6.7 Simplified sketch of retailer– fleet operator–EV customer integration.

- Assuming the retailer and the fleet operator are operating by plan, the surplus flexibility of the fleet operator can be used by the retailer to participate in the ancillary service markets. This scenario will be in favour of the retailer, fleet operator and society, since this will present a helpful tool in integrating more fluctuating wind energy.
- Owing to the flexibility aggregated by the fleet operator, this charging strategy can be used for some kind of future smart grid services. One example of this could be services to the DSO; for example, congestion management.
- In control theory terms, fleet-operator-based charging performs closed-loop control, which in control quality terms is better than open-loop control (as in the case of price signal strategy).

It should be noted that for simplicity the retailer box in Figure 6.6 consists of both retailer and wholesaler. In reality, it will be either the retailer or the wholesaler – depending on the skills of the retailer – who has a contract with the fleet operator. A simplified sketch of retailer (wholesaler)–fleet operator and fleet operator–EV customer interrelation is presented in Figure 6.7. In brief, the flexibility of the EVs is aggregated and served by the fleet operator based on contractual terms. The aggregated flexibility is further aggregated by the retailer (wholesaler), who does the market operation. Acquired energy is communicated back to the fleet operator, who will serve the EV customers with charging.

The set-up presented in Figure 6.7 can, in principle, relatively easily be extended to the intraday market and also the power regulation market without any changes in the current market configurations. As a result, the competition in these markets will increase in benefit for all customers.

The 'power dispatch' box sketched in Figure 6.7 illustrates a dispatch controller with the planned charging as reference (input) signal and with the charging signals for each grid-connected EV as output signal. Thus, the objective of the dispatch controller is to keep track of all the grid-connected EVs and make sure that all the EVs are served according to the contracts; that is, no EV may enter a situation where the contractual state of charge has not been reached when the EV is scheduled to be disconnected.

The charging plan is based on historical data for each EV combined with expectations (special events, weather data, etc.) for the coming day. However, there is always a risk that the plan and reality will diverge. Owing to the flexibility of the EVs, the dispatch controller can rearrange the charging within its EV portfolio; that is, as long as the contractual states of charge are not violated. Here, two extreme scenarios exist: either the plan overestimates reality or it underestimates reality. Anyway, either scenario will result in lack of flexibility; the EVs will be fully charged too early or, even worse, the charging of the EVs will not keep up with the contractual obligations. Thus, the dispatch controller must possess a kind of supervision algorithm in order to catch these scenarios well in advance. Nevertheless, whenever one of the two scenarios is identified, the solution is rather simple. If the plan overestimates reality, the fleet operator should ask the retailer to sell surplus energy in the intraday market or put corresponding bids into the power regulation market. If the plan underestimates reality, the fleet operator should ask the retailer to buy more energy in the intraday market or put corresponding bids into the power regulation market. If the acquired surplus energy is bought at a spot price, which could be expected to be better than the coming day's spot price, the fleet operator could also decide to do nothing and to reduce the next day's energy purchase equivalent to the surplus energy.

As a small add on to the previous paragraph, it should be noted that, if the EV portfolio is sufficiently large, the fleet operator can use its surplus flexibility to provide service in the intraday market as well as the power regulation market. The main conditions for this would be good abilities in forecasting the flexibility of the EV portfolio and a good dispatch controller.

The fleet operator/retailer set-up described above is more or less embodied in the cooperation between Better Place and DONG Energy [5]. Since the beginning of 2012, potential EV customers have been able to close contracts with Better Place.

6.3   Case Studies of EV Fleet Management

In this section, simulations of five charging methods (possible implementations) are carried out covering three of the charging strategies: dumb, timer-based and fleet operator based charging. The charging methods, which are named: dumb charge, dumb charge home, timer, fleet and fleet optimal, are compared and benchmarked against each other using the spot price as cost function. Two important parameters, the number of EVs and the maximum charging power, are investigated and some general conclusions are drawn regarding the consequences and the possible usage of each of the charging methods.

6.3.1   System Description

The influence of different penetration levels of EVs in the fleet on Bornholm is analysed using the five different charging methods. The number of vehicles is chosen to be 2000 and 4000, corresponding to penetration levels of 10% and 20% shares of the car population respectively. Also, the power level of the grid connection available to the EVs is of importance when analysing grid impact. Therefore, the different impacts of one- or three-phase charging are also analysed.

Energy Planning Tool

A simulation program is implemented in MATLAB® in order to analyse the consequences of the different charging methods. The main components in the simulation code are described in the following.

The input data to the simulation program are:

1. driving data containing the times when the individual vehicles connect to and disconnect from the grid, as well as the number of kilometres driven;
2. the average spot price on an hourly basis;
3. basic scenario data – for example, the number of vehicles, the available number of phases, the current for each phase, battery capacity.

The main program (see Figure 6.8) processes the input data and calls the functions for each of the charging methods, which return the charging plans for each vehicle. For some of the charging tools, additional parameters are required, such as starting time for the 'timer tool'. The result of the computation is thus a plan for each individual vehicle.

After the individual charging plans have been computed they are post-processed to calculate the total power consumption for each of the primary substations on the island as well as the total charging power for the whole EV fleet.

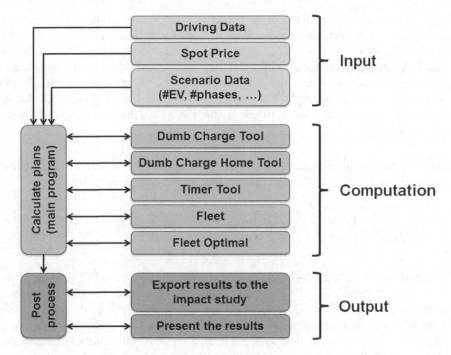

Figure 6.8   Overview of the program implemented for the simulations.

It is assumed that the particular vehicle will be acting like it did according to the reported driving pattern in the National Travel Survey [6]. Even though one might expect large variations in the driving pattern from day to day for each vehicle (an aspect which is not included in the National Travel Survey), the average driving pattern of a large number of vehicle for one day is expected to be representative for driving pattern for the next day.

As discussed in Section 6.2.2, the limit of the battery charge has been limited to 85% of the maximum battery capacity to increase battery life. To simplify the terminology, 'full SOC' refers to the chosen limit, which in this case is 85% of the maximum charge of the battery. It should be noted that the conclusions of the simulation results do not depend on the chosen limit of the batteries.

Each of the charging methods is described in the following.

- In the 'dumb charge' method, each EV is charged up to full SOC each time it is connected to the grid.
- In the 'dumb charge home' method, each EV is charged up to full SOC, when it returns home after the last driving tour of the day. The driving data set contains a purpose field describing the purpose of the driving tour, which is used in order to determine when the EV is driving home. Data when the EV is not driving home as the last tour of the day are not included in the analysis.
- The 'timer' method is configured so the EV starts charging at a certain time of the day (22:00), if it is connected. If not connected, it will start charging when it returns back home after 22:00 and will charge up to full SOC, if not disconnected before.
- The 'fleet' method is designed so the EV will be optimally charged up to full SOC every time the EV is connected to the grid, and also during the day when the cost of electricity

is high. This method will tend to put the charging in the middle of the day when the price of electricity slightly decreases compared with the two peak hours (one in the morning and one in the afternoon).

- The 'fleet optimal' method is the optimal charging method, only charging when the expected spot price is at its lowest. The assumption here is that the battery capacity is large enough to support the total driving of the following day. Thus, the charging is placed during the time when the expected spot price is low – which in a Danish context typically is during night-time.

An optimization problem is solved for both the 'fleet' and the 'fleet optimal' methods as given below. For each of the EVs, the following optimization problem is solved:

$$E = \min \left\{ \sum_i (\text{sp}_i \cdot \dot{W}_{\text{el},i}), \quad \sum_i \dot{W}_{\text{el},i} \cdot T_s = Q \quad \text{and} \quad \dot{W}_{\text{el},i} \leq \dot{W}_{\max} \right\}$$

where E (euros) is the energy cost for charging the current EV, sp_i (€/kWh) is the expected spot price at time sample i, $\dot{W}_{\text{el},i}$ (Wh) is the charging power for the current EV at time sample i, T_s is the sample time (900 s), Q is the total required energy to charge the battery from the initial conditions to the contractual state of charge and $\dot{W}_{\max}$ (W) is the maximum possible charging power for the EV. The time sample i goes from one to the number of time samples. In the case of 'fleet' the charging time is limited to the time between two drives. In the case of 'fleet optimal', the charging time is any interval when the EV is connected. The solution of the equation is a vector $\dot{W}_{\text{el},i}$ which satisfies that the cost for charging is the lowest possible under the constraints that the required charge must be supplied and the maximum possible charging power may not be exceeded.

Driving Data

Special attention has been paid to the driving pattern. The overall assumption is that the driving pattern of the EVs on Bornholm is not particularly different from the overall driving pattern reported in the National Travel Survey 2006–2010 for ordinary gasoline and diesel cars [7]. Figure 6.9 presents the accumulated frequency function of vehicles driving less than a certain number of kilometres a day.

In order to find the percentage of vehicles driving less than or equal to, for example, 130 km/day, follow the line at 130 km/day on the x-axis and find the result 0.92 corresponding to 92% on the y-axis.

The driving data are thoroughly analysed and the main conclusions are given below. Some data have been sorted out as irrelevant for this analysis. All vehicles with extensive driving (>130 km/day) are left out of this analysis as they are unlikely candidates for electrical transportation given the present range limitations of available EVs. Some preliminary observations are that:

1. Most drivers cover distances in a day for which electrical vehicles may be used (92% drive less than 130 km/day; refer to Figure 6.9).

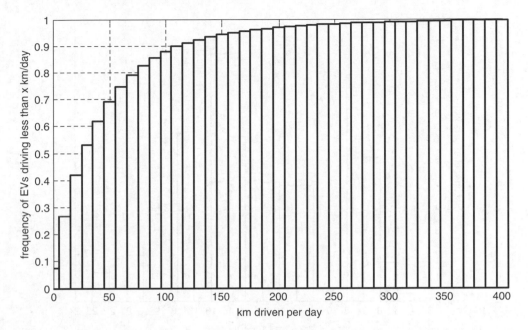

Figure 6.9 Accumulated frequency function of vehicles driving less than a certain number of kilometres a day.

2. Most of the vehicles are standing still, and may thus in the case of EVs be connected to the grid most of the time – on average, 94.2% of the day; please refer to Figure 6.10.
3. The influences of the day of the week (weekday, Saturday or Sunday) and the area (city, town or rural area) are not important in this study (please refer to the detailed analysis [2]). Average data over all days of the week, as well as in all areas, will be used in the following analysis of the charging methods and the impact studies.

Based on the driving data, the availability of EVs can be derived as the number of standing/parked cars, as presented in Figure 6.10, being grid disconnected.

It can be seen that the vehicles are parked most of the time of the day. Assuming EVs to behave like gasoline cars in the survey, it is concluded that EVs may possibly be connected to the grid most of the time of the day.

Spot Price

The spot price is used in the analysis to determine the cost of charging at the different time intervals. Thus, the results of the analysis depend largely on the spot price time-series used. In this analysis, the hourly average spot price was used. The hourly average spot price is calculated as the average price at each hour of the day in the price area DK2 (area of Zealand and Bornholm, etc.) for 2009, as shown in Figure 6.11. Also, the standard deviations from the average value are shown, giving some indication of the variation of the spot price. The spot

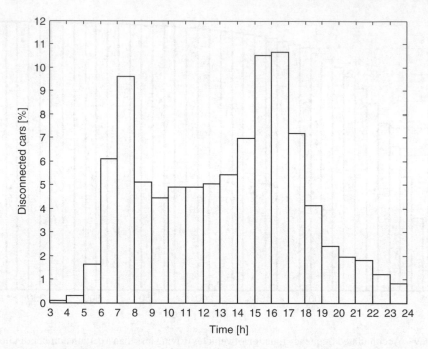

Figure 6.10 The percentage of disconnected cars on daily basis. Data from before 03:00 are not available.

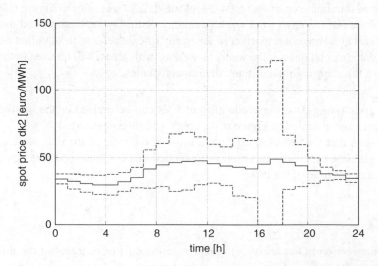

Figure 6.11 Average spot price (solid line, €/MWh) in DK2 (Zeeland and Bornholm, etc.) in 2009 as well as the spot price plus/minus one standard deviation (dotted lines).

Table 6.1 Overview of load cases considered.

Load case	No. of EVs	No. of phases	Current (A)	Max. charging power (kW)	EV penetration (%)
1	2000	1	16	3,7	10
2	2000	3	16	11,0	10
3	4000	1	16	3,7	20

price is seen to have two peaks during the day: one at 11:00 and one at 17:00. The lowest spot price occurs at 03:00–05:00.

The spot price is assumed to be independent of the consumption. This might not be fully true when the charging power from the EVs is comparable to the overall consumption. If the EVs are charged during night-time and the charging power is significant compared with the overall consumption, then electricity cost will tend to increase during night-time. This effect due to the free market forces will occur when the charging power is significant compared with the overall electricity consumption. In the present work some small effect of this might be present, but it is not expected to change the results significantly.

For simplicity, the average spot price will be used as expected spot price in the energy planning tool described in Section 6.3.2.

Historical market data are available as a download from NordPool Spot (www.nordpoolspot.com).

6.3.2 Scenario Description

Three load cases are computed and the influence of varying the number of EVs (the penetration of the EVs of the total fleet of vehicles) and the number of phases (determining the maximum charging power) are investigated. For simplicity, the capacity of one battery (for all EVs) is 23.3 kWh. The scenarios investigated are shown in Table 6.1.

These load cases are used for the scenarios listed in Table 6.2; that is, load case 1 will be discussed first and load cases 1 and 3 will afterwards be compared in order to determine how EV penetration and the number of phases available influence the load pattern.

These scenarios will be benchmarked against the five dedicated charging methods, which have been derived from the general strategies described in Section 6.2.3.

Table 6.2 Overview of scenarios considered.

Scenario	Description	Load case
1	Comparing charging methods for 2000 EVs one-phase charging 16 A	1
2	Comparing 2000 and 4000 EVs	1 + 3
3	Comparing one- and three-phase charging	1 + 2
4	Limiting maximum power consumption	1

Scenario 1: Comparing Charging Methods for 2000 EVs with One-phase Charging 16 A

The simulation results for 2000 EVs all charging with one-phase 16 A are shown in the following. Figure 6.12 shows the total charging power for all 2000 EVs as a function of the time of the day for each of the five charging methods. In order to be able to understand the performance of the fleet methods, the spot price has been included in Figure 6.12.

The 'dumb charge' method is seen to have two peaks per day corresponding to the time of the day when most people drive to and from work. The peaks coincide with the peak hours of the spot price and the 'dumb charge' method is consequently also the most cost-expensive charging method with an average of 44.2 €/MWh; see Table 6.3. The average cost per megawatt-hour using the different charging methods is shown in Figure 6.13. In the 'dumb charge home' strategy, the EVs are only charged at home, thus giving a peak at about 17:00,

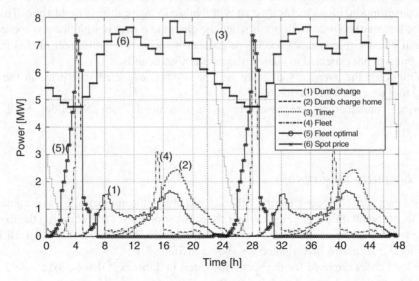

Figure 6.12   Aggregated charging power on the 60 kV grid as function of the time of day for each of the charging methods. Sample time is 15 min. 2000 EVs, one-phase charging, 16 A.

Table 6.3   Average cost for charging the EVs using each of the charging methods of one-phase charging, 16 A.

Charging method	Average cost (€/MWh)
Dumb charge	44.2
Dumb charge home	43.9
Timer	35.4
Fleet	36.0
Fleet optimal	30.1
Maximum spot price	49.0
Minimum spot price	29.6

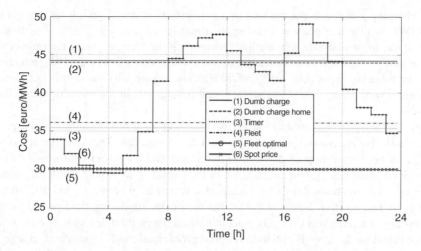

Figure 6.13   Average cost per megawatt-hour for each charging method compared to the spot price.

when people return from work. This type of charging will prevail until we have extended the publicly available charging posts. The 'fleet' method also has two peaks per day. However, these peaks correspond to the time of day where the price is lowest. Thus, the 'fleet' provides charging all day, but the highest share of charging is provided during low-cost hours. The 'fleet optimal' method is based on the assumption that charging only need to be performed once a day and, thus, may be placed at the cheapest time interval at night. This results in a peak during the least expensive price.

The simulation results in Figure 6.12 are directly comparable, in the sense that the energy served (area below the curves) is identical. In other words, the five different charging methods all manage to fulfil the driving needs of all the EVs. This fact really illustrates the flexibility of the EVs; namely, the time of charging can be moved around without interfering with the main purpose of the EVs – to do transport whenever needed.

The spot price as function of the time of the day is also shown, giving some insight into the performance of the charging methods. For example, the 'fleet optimal' method is seen to give an average cost almost the same as the lowest possible spot price, whereas the 'dumb charge' gives an average cost at the upper end of the spot price.

A simple extension of this method is the 'timer'-based method. In this case, charging is postponed for all EVs until 22:00, giving an instant change in power consumption from 0 to 7.3 MW. The time at which the charging is initiated is here somewhat arbitrarily chosen. A well-chosen time could show that the timer method would perform as well as the 'fleet optimal' method. This is in reality *not* the case, which would be clear if we were to calculate the cost using the actual spot price each day of a year and not an average spot price. This analysis is omitted here. Even though this rapid change in power consumption is not a desired behaviour seen from a grid stability point of view, the average cost of 35.4 €/MWh is quite good compared with the small amount of effort. Using the 'fleet' method, the EVs will be charged each time they are connected to the grid (all time intervals when not driving). The charging is optimal, in the sense that the EVs are charged at the most favourable time when parked and the EVs are always charged as much as possible each time they are connected.

But the charging scheme is not optimal in a cost-effective sense, giving an average cost of 36.0 €/MWh, which is slightly worse than the timer-based method. The final method, 'fleet optimal' is the most cost-effective method, postponing the charging until the time of the day when the spot price is at its minimum, giving an average cost of 30.1 €/MWh, which is very close to the minimum spot price of 29.6 €/MWh. The reason why the cost is not identical to the minimum spot price is that the charging time is longer than the timeslot where the price was lowest.

Comparing the power consumption of the charging methods, the 'dumb charge' looks attractive as the power consumption is spread out during the day. The disadvantage is that the charging is done when the spot price is at its maximum and the power consumption thus adds to the peak consumption, giving an increase in the overall grid impact. Both the 'timer' and the 'fleet optimal' methods show very high peaks, which might have an undesired impact on grid stability. The positive aspect of this is that the peaks are during time periods where the other consumption (all except EVs) is at its minimum. Even though giving rise to some high peaks, the effect on the overall grid load is not as problematic as for the 'dumb charge'.

Scenario 2: Comparing 2000 and 4000 EVs

One of the important parameters to investigate is the number of EVs or, consequently, the penetration of EVs in the total fleet of vehicles. Changing the number of EVs from 2000 to 4000 – both using one-phase 16 A charging – roughly changes the penetration from 10% to 20% on Bornholm; see Figure 6.14. The figure shows that the maximum power consumption scales directly proportionally with the number of EVs.

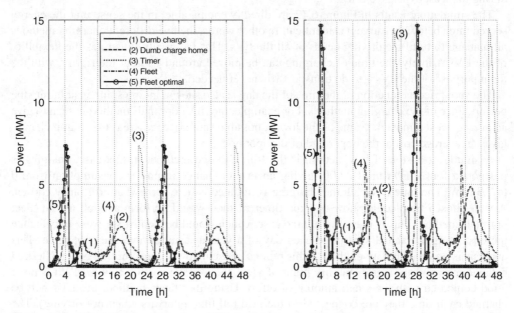

Figure 6.14 The influence of number of EVs on the power consumption – aggregated power consumption on the 60 kV grid.

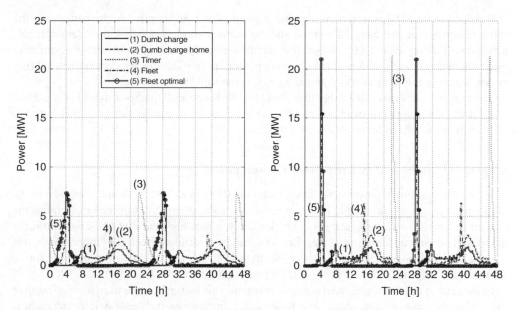

Figure 6.15 Aggregated power consumption on the 60 kV grid for 2000 EVs 16 A using one-phase charging (left) and three-phase charging (right).

Scenario 3: Comparing One- and Three-phase Charging

Another important parameter to investigate is the number of phases available for charging; see Figure 6.15. Assuming a resistive load (i.e. $\cos(\phi) = 1$), the charging power maxima for a single EV using one- and three-phase charging are:

$$\text{maximum power (one phase)} = \text{one phase} \times 16 \text{ A} \times 230 \text{ V} = 3680 \text{ W}$$
$$\text{maximum power (three phase)} = \text{three phase} \times 16 \text{ A} \times 230 \text{ V} = 11\,040 \text{ W}$$

Going from one-phase to three-phase charging decreases the charging time. From Figure 6.15 it is seen that the influence is very different depending on the charging method. There is hardly any difference for the 'dumb-charge' method as the charging consumption is spread throughout the day. Comparing the two cases for the 'timer' and the 'fleet optimal' methods shows that the maximum power consumption is tripled. This is due to the fact that all EVs are charging at the same time. An increase in maximum power of a factor of 3 also gives a factor of 3 on the power consumption in those few time intervals. The 'fleet' method is somewhere between the other two cases, with a doubling in power consumption when going from one to three phases.

It should also be noted that the flexibility using three-phase charging is greater than when using one-phase charging. The reason is that since the charging time using three-phase charging is roughly one-third of the charging time using one-phase charging, then the freedom to choose when the charging power is placed is higher.

Finally, it should be noticed that the total amount of energy supplied to the batteries is the same regardless of the charging method and the number of phases. Thus, the 'fleet optimal' method will increase the charging power to the maximum limit at the most cost-efficient time intervals, narrowing the shape of the charging power curve – but the area below the charging curve, which is the same as the supplied energy, is the same in both cases.

The conclusion is that at least the 'timer' and the 'fleet optimal' methods should be modified to limit the maximum power in order to respect the maximum load of the grid on Bornholm. This problem is solved in the following subsection.

Scenario 4: Limiting Maximum Power Consumption

Limiting the maximum power is very easy using the 'fleet' and 'fleet optimal' methods. In the mathematical model one simply adds a condition for the optimization problem limiting the maximum power to a user-specified value. Similarly, a real-life fleet operator could limit the maximum power consumption of their fleet. To achieve a similar effect, the 'timer' method is modified in such a way that some EVs are set to start earlier (up to 1 h) and some later (up to 1 h) than the originally chosen time of 22:00 (see Figure 6.16).

In the case of the 'fleet' and 'fleet optimal' methods, this feature may be used to help stabilize the grid in the case of bottlenecks (grid limitations). In this way the fleet operator may adjust the power consumption to the available capacity in the grid and thus both increase the ability to regulate the overall power balance and also, on a longer time scale, help postpone/prevent new costly extensions of the grid. The economic quantification of this benefit is somewhat difficult to estimate owing to the large number of uncertainties, so the statement here is that it has some value, which is of importance in the overall evaluation.

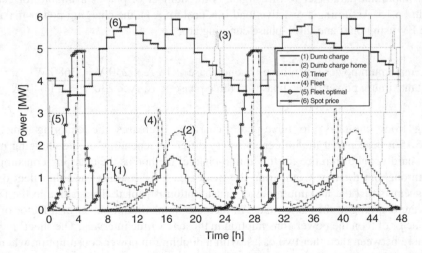

Figure 6.16 Aggregated power consumption on the 60 kV grid when the maximum power for the 'fleet' and 'fleet optimal' methods have been limited to 5 MW (chosen by the fleet management operator). The EVs using the 'timer' method start charging somewhat displaced in time (some before and some after 22:00).

6.3.3　Conclusions on the Case Studies of the Charging Methods

Five charging methods have been investigated: 'dumb charge', 'dumb charge home', 'timer', 'fleet' and 'fleet optimal'.

- 'Dumb charge' is the simple method where the EV is charged each time it is parked. This is the simplest method, but also the most costly using the spot price as cost function. Even though the curves of the charging power look smooth compared with the other, this method will cause the bulk power consumption to be in the peak hours, thus giving a possible high impact on the grid.
- The 'dumb charge home' method is also the simple method where the EV is charged up to the maximum limit each time it is parked at home. This method will prevail until we have installed an extensive system of public charging posts. This is also the charging method which gives the highest impact on the grid system, as the main power consumption is concentrated in the evening peak hours and thus gives a higher cost for charging when compared with the other methods.
- The 'timer'-based method is equivalent to 'dumb charge', but limited so that charging starts after peak hours. In this analysis the starting time was set to 22:00. This gives a high peak in the charging power just at 22:00 which can be smoothed out by starting the charging of some EVs before and others after 22:00. This method is quite simple and gives a remarkable low cost when using average prices.
- 'Fleet' is an optimal method in the sense that it chooses the cheapest time interval to charge the EV to maximum charge each time the EV is connected to the grid (i.e. parked). As the EVs are used throughout the day, a significant part of the charging will also be during the day with high spot prices. Thus, this strategy is not very cost efficient even though it gives a high security for the driver concerning charge on the batteries (but not as high as the 'dumb charge').
- 'Fleet optimal' is the cost-optimal method as it chooses the most cost-efficient time intervals during the night for charging the EV. This method has two major benefits compared with the others. One is the ability to adjust the power consumptions to the time intervals with lowest spot price, giving a clear economic advantage. The other is the ability to adjust the power consumption to the current available capacity in the grid and thus help avoiding bottleneck problems. The latter benefit may help preventing new costly extensions of the grid system.

It has been observed that all the charging methods can cause load peaks. This will sooner or later cause problems in the distribution grid. The fleet operator has the ability to limit its consumption, both generally and in specific local areas. With the current market set-up, there are no incentives for the fleet operator to do so. Anyway, information about such limitations will have to be conveyed by the DSO – one possibility could be a future DSO congestion market.

One assumption in this study was a negligible feedback from the consumption of the EV portfolio population on energy prices. This might be true in the case of Bornholm, but certainly not on a national scale. A significant-sized EV portfolio would, of course, influence the spot price. Since the fleet operator would try to allocate as much charging as possible during low-cost hours, the cost of these low-cost hours would increase with increasing charging allocation in these hours. However, this cost increase would, of course, be smaller than an equivalent charge allocation in medium- or even high-cost hours. The market forces will try to fit-in additional flexible energy consumption at times with low prices – which in a Danish context normally would be equivalent to hours with high penetration of wind power. This is a very beneficial correlation – except when the low-cost hours occur during the high peak hours around lunch and dinner times. When this happens, the DSO congestion market becomes even more important.

6.3.4   Future Implications

The congestion market is supported by the economic interests of the DSO and the consumers. Introducing a congestion market, the DSO may postpone or omit new grid reinforcements. Using the grid to the limit may also increase the risks of getting blackouts and expenses for getting the grid line up working again (changing fuses). The costs of these expenses are difficult to estimate as they depend on the local circumstances. It should be noted, however, that particularly the expenses for grid reinforcements are significant. The above expenses are associated with the DSO and may be decreased, introducing a congestion market. For the consumers, a congestion market introduces limits on the overall maximum power. Thus, the high peaks resulting from charging EVs in the hour with the lowest spot price must be spread out over some hours (but still with low spot price). This cost should be counterbalanced by the reduced costs for the DSO.

References

[1] Hay, C., Togeby, M., Bang, N.C. *et al.* (eds) (2010) Introducing electric vehicles into the current electricity markets, http://www.edison-net.dk/Dissemination/Reports/Report_004.aspx (accessed January 2013))
[2] Wu, Q., Jensen, J.M., Hansen, L.H. *et al.* (2012) EV Portfolio Management and Grid Impact Study. Edison report.
[3] Bjerre, A. (2010) The Edison picture, http://www.edison-net.dk/Dissemination/Reports/Report_003.aspx.
[4] Bang, C., Fock, F. and Togeby, M. (2011) Design of a real time market for regulating power, FlexPower WP1 – Report 3, http://www.ea-energianalyse.dk/reports/1027_design_of_a_real_time_market_for_regulating_power.pdf (accessed January 2013).
[5] Copenhagen Capacity (2009) Fact sheet – transportation and energy storage, http://www.copcap.com/media/1033_604.pdf (accessed January 2013).
[6] Christensen, L. (2011) Electric vehicles and the customers, http://www.edison-net.dk/Dissemination/Reports/Report_011.aspx (accessed January 2013).
[7] Transportvaneundersøgelsen (2011) Data of The Danish National Travel Survey.

7

Analysis of Regulating Power from EVs

Qiuwei Wu, Arne Hejde Nielsen, Jacob Østergaard and Yi Ding
Centre for Electric Power and Energy (CEE), Department of Electrical Engineering, Technical University of Denmark, Lyngby, Denmark

7.1 Introduction

With more and more renewable energy integrated into power systems, it has become an attractive option to use electric vehicles (EVs) to balance the uncertainties introduced by renewable energy.

The possibility of using vehicle to grid (V2G) to improve wind power integration was studied by Brassin [1]. The traffic data were used to calculate the vehicle fleet availability. It was concluded that it is possible to have EVs providing instantaneous disturbance and manual reserve to help integrate more wind power. The feasibility study of implementing V2G scenario in Denmark was done by Chandrashekhara *et al.* [2]. The system constraints for integrating EVs into power systems were examined and the technical and economical viabilities of various possible V2G architectures were studied. It was concluded that the V2G technology can assist in realizing the Danish government goal of '50% of the total energy consumption to be met by wind power in 2020'. A V2G demonstration project was implemented in AC Propulsion, Inc. to evaluate the feasibility and practicality of EVs providing regulation service [3]. A test vehicle was fitted with a bidirectional grid power interface and wireless internet connectivity to carry out the demonstration. It was shown that it is feasible for EVs to provide regulation service from a technical and economical point of view.

Three types of EVs were studied by Kempton and Tomic [4] to check the feasibility of the V2G concept from an economic perspective and four markets were used to carry out the quantitative analysis of the economic return for EV users by providing a V2G service. It was concluded that EVs should participate in the ancillary service market to be profitable by providing an ancillary service. Andersson *et al.* [5] showed under what conditions plug-in

Grid Integration of Electric Vehicles in Open Electricity Markets, First Edition. Edited by Qiuwei Wu.
© 2013 John Wiley & Sons, Ltd. Published 2013 by John Wiley & Sons, Ltd.

hybrid electric vehicles can generate revenues. The actual data for four months in 2008 in Germany and Sweden were used to do the analysis.

In this chapter, the potential of using EVs to provide regulating power in Denmark was investigated using the driving pattern data and the regulating power market data to obtain the regulating power capacity from EVs and the economic return for EV users.

7.2 Driving Pattern Analysis for EV Grid Integration

In order to facilitate the analysis of EV integration into the Danish power system, the driving data in Denmark were analysed to extract the information of driving distances and driving time periods which were used to represent the driving requirements and the EV unavailability. The Danish National Transport Survey data (TU data) were used to carry out the driving data analysis. The average, minimum and maximum driving distances were obtained for weekdays, weekends and holidays to illustrate the EV users' driving requirements on different days. The EV availability data were obtained from the driving time periods to show how many cars are available for charging and discharging in each time period. The statistical analysis software SAS was used to analyse the driving data.

7.2.1 Driving Distance Analysis

The driving distance information is very important and reflects the driving requirements of EV users. It can be used to determine the battery size for different kinds of customers.

The driving distance data from the TU data were analysed to get the average driving distance and driving distance distribution for different days.

The average driving distance data are listed in Table 7.1. The overall average daily driving distance is 40 km. For Mondays, the average driving distance is 43.399 km. The average driving distances for Saturdays and Sundays are 34.074 km and 29.723 km, respectively.

The driving distance distributions are illustrated in Figure 7.1. It is shown that about 70.41% of car users drive equal or less than 40 km. The driving distance distribution on Mondays is illustrated in Table 7.2 and Figure 7.2. The individual and cumulative driving distance distributions for weekends are illustrated in Table 7.3 and Figure 7.3 and Figure 7.4

The driving data analysis shows that the average diving distance is shorter on the weekends than on weekdays and the average driving distance is shortest on Sundays.

7.2.2 EV Availability Analysis

In the TU data, the starting and ending times of all parking time periods from one respondent were combined to determine the time periods which are available for EV charging and discharging. The survey number and day were used to get all the parking data for one respondent.

Table 7.1 Average driving distance.

Day type	Average driving distance (km)
All days	40
Monday	43.399
Saturday	34.074
Sunday	29.723

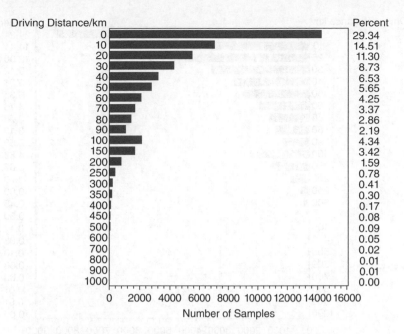

Figure 7.1   Average driving distance distribution on all days.

Table 7.2   Average driving distance distribution on Mondays.

Driving distance (km)	Individual (%)
0	24.21
10	14.41
20	12.00
30	9.42
40	7.21
50	6.35
60	4.76
70	3.78
80	3.13
90	2.50
100	4.88
150	3.67
200	1.70
250	0.80
300	0.43
350	0.30
400	0.17
450	0.08
500	0.10
600	0.06
700	0.03
800	0.01
900	0.01
1000	0.00

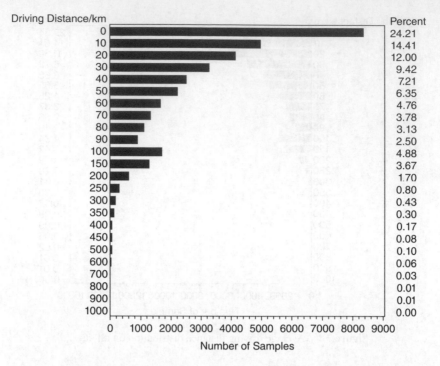

Figure 7.2 Average driving distance distribution on Mondays.

Table 7.3 Average driving distance distribution on weekends.

Driving distance (km)	Saturday individual (%)	Sunday individual (%)
0	37.57	45.51
10	15.11	14.40
20	10.14	9.11
30	7.64	6.55
40	5.27	4.51
50	4.67	3.35
60	3.51	2.57
70	2.61	2.17
80	2.36	2.05
90	1.27	1.60
100	3.36	2.76
150	3.18	2.49
200	1.46	1.18
250	0.78	0.70
300	0.48	0.25
350	0.25	0.36
400	0.15	0.21
450	0.10	0.05
500	0.07	0.09
600	0.3	0.07
700	0	0.01
800	0	0.01
900	0	0
1000	0	0

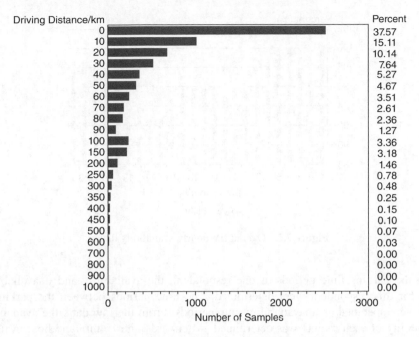

Figure 7.3 Individual driving distance distribution of Saturdays.

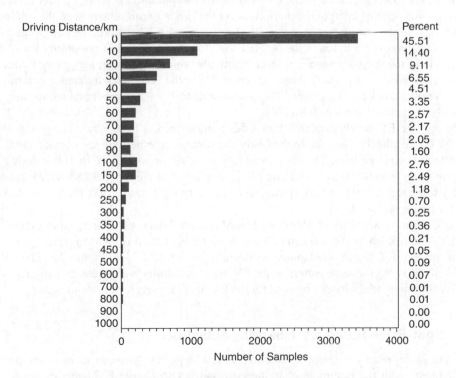

Figure 7.4 Individual driving distance distribution of Sundays.

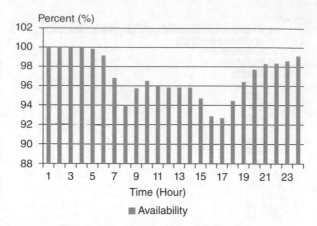

Figure 7.5   Overall EV hourly availability data.

From the parking time periods of one respondent, the availability and unavailability of the EV for this respondent were determined. The time periods between the parking time periods were specified as unavailability time periods. From the raw data, the availability or unavailability of each minute was determined. Afterwards, the 15 min and hour availability data were calculated based on the minute data.

For the EV grid integration study, EV demand will be put into the energy power market to determine the optimal charging scenario to meet the driving requirement with the minimum energy cost. In the energy power market, the time step is 1 h. Therefore, the hour availability data were determined. On top of the optimal charging study, there is a possibility for EVs to participate in the ancillary service market to provide regulating power, secondary frequency control (DK West) and primary frequency control. For the secondary frequency control, the operation time can be up to 15 min. The quarter availability data were obtained to carry out the study of ancillary services from EVs.

The average EV hourly availability of 1 day is illustrated in Figure 7.5. The results show that the EV availability is quite high if only the driving time periods are considered as the unavailability time periods. The EV availability is 100% or very close to 100% during the early morning from 00:00 to 06:00. The EV availability is in the range 96.83–92.73% during the day time from 07:00 to 19:00. During night-time, from 20:00 to 24:00, the EV availability is in the range 97.73–99.14%.

The EV hour availability of Mondays, Saturdays and Sundays is illustrated in Figure 7.6. The EV hour availability on weekends is a little bit higher than it is on Mondays.

The average EV 15 min availability is illustrated in Figure 7.7 and Figure 7.8. The EV 15 min availability has the same pattern as the EV hour availability with fluctuation in each hour. The EV 15 min availability can be used for the EV ancillary service provision study.

7.3 Spot-Price-Based EV Charging Schedule

The spot-price-based EV charging schedule is to charge EV batteries to meet the driving requirements with the minimum charging cost and not to charge EV batteries during the

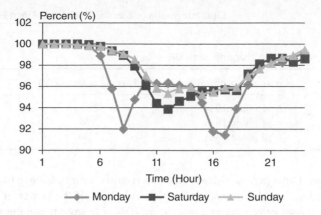

Figure 7.6 Hourly availability data of Mondays, Saturdays and Sundays.

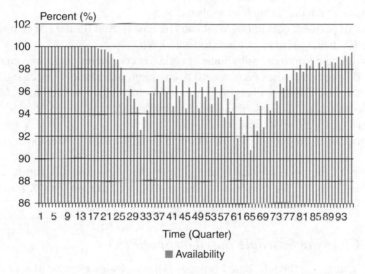

Figure 7.7 Average EV 15 min availability data.

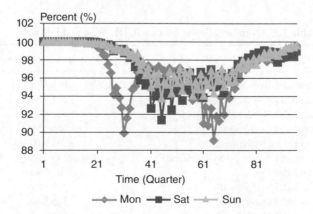

Figure 7.8 Average EV 15 min availability data on Mondays, Saturdays and Sundays.

Table 7.4 Proposed EV home charging methods.

Charging method	Charging specification	Charging power (kW)	Charging time (h)
Method 1	1 phase 10 A	2.3	5.65
Method 2	1 phase 16 A	3.68	3.53
Method 3	3 phase 16 A	11.04	1.18

normal high-demand time periods. Ideally, the power grid constraints have to be considered in the charging schedule in order to respect the physical limits of the power grid. However, the intention of the analysis of regulating power from EVs is to investigate the potential of how much regulating power capacity is available from EVs and how much the economic return will be by having EVs provide regulating power for the system. Therefore, the power grid constraints are not included in such an analysis.

The number of personal cars in DK West and DK East is used for the EV charging schedule study and the EV battery size is assumed to be 25 kWh.

For the EV charging options, only home charging is considered and three charging power levels are used in the analysis: one phase 10 A charging, one phase 16 A charging and three phase 16 A charging. The three charging power levels are defined as Method 1, Method 2 and Method 3 illustrated in Table 7.4. The charging power is 2.3 kW for Method 1, 3.68 kW for Method 2 and 11.04 kW for Method 3.

The charging times are 5.65 h, 3.53 h and 1.18 h for Method 1, Method 2 and Method 3 respectively.

The numbers of personal car were obtained from Danmarks Statistik (http://www.dst.dk/) and are listed in Table 7.5. The numbers of personal cars are 1 206 641 in DK East and 888 494 in DK West. These two numbers were used for the EV charging study.

7.3.1 EV Charging Schedule Based on Spot Price

The study of charging schedule based on spot price investigates what the system demand will be if all the customers choose to charge their EVs during the expected low spot-price time periods.

Table 7.5 Number of personal cars in DK West and DK East

Area	Personal car number
Landsdel København by	165 873
Landsdel Københavns Omegn	201 789
Landsdel Nordsjælland	189 870
Landsdel Bornholm	16 358
Landsdel Østsjælland	91 546
Landsdel Vest – og Sydsjælland	239 416
Landsdel Fyn	188 590
Landsdel Sydjylland	294 809
Landsdel Østjylland	316 244
Landsdel Vestjylland	176 763
Landsdel Nordjylland	230 035

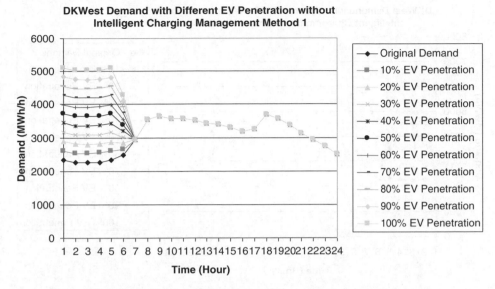

Figure 7.9   DK West system demand with different EV penetration for Method 1.

The DK West and DK East system demands with different EV penetration for Method 1 are illustrated in Figure 7.9 and Figure 7.10 respectively. For charging Method 1, the charging time is 5.65 h and the customers choose to charge EVs from 00:00 to 06:00, which are the expected 6 h of lowest price.

The DK West and DK East system demands with different EV penetration for Method 2 are illustrated in Figure 7.11 and Figure 7.12 respectively. For charging Method 2, the charging

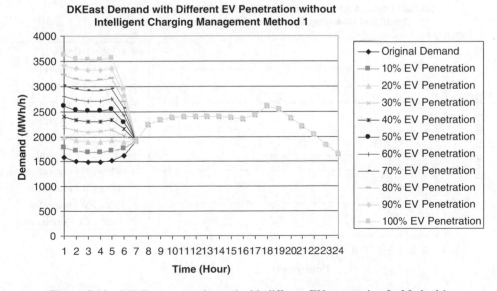

Figure 7.10   DK East system demand with different EV penetration for Method 1.

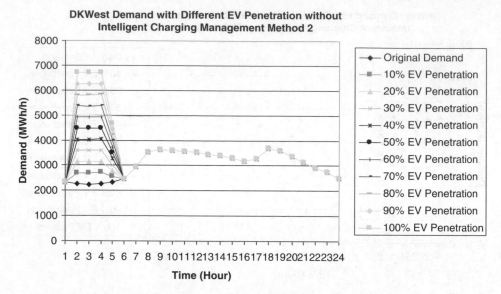

Figure 7.11   DK West system demand with different EV penetration for Method 2.

time is 3.53 h and the customers choose to charge EVs from 00:00 to 04:00, which are the expected 4 h of lowest prices.

The DK West and DK East system demands with different EV penetration for Method 3 are illustrated in Figure 7.13 and Figure 7.14 respectively. For charging Method 3, the charging time is 1.18 h and the customers choose to charge EVs from 01:00 to 03:00, which are the normal 2 h of lowest prices.

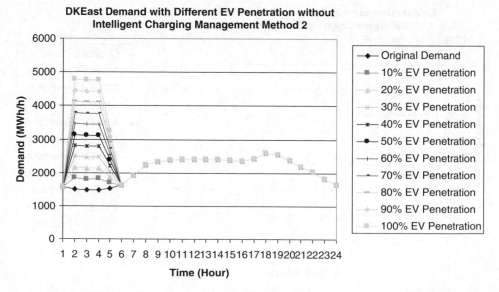

Figure 7.12   DK East system demand with different EV penetration for Method 2.

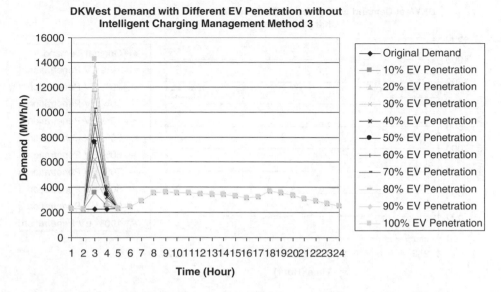

Figure 7.13   DK West system demand with different EV penetration for Method 3.

The system demands with EV charging for Method 1, Method 2 and Method 3 show that the extra demand from EV charging will create another demand peak, especially for Method 3. Approximately, the EV penetration levels for Method 1, Method 2 and Method 3 are 50%, 35% and 12% respectively without putting extra stress on the existing system. The criterion for determining the penetration limits is that the new peak demand should not be higher than the original one.

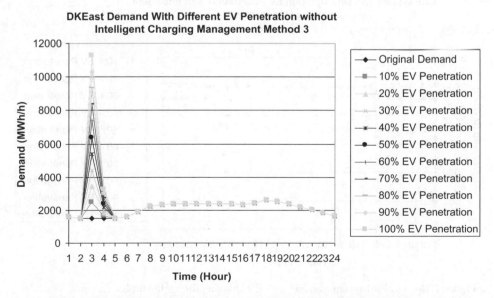

Figure 7.14   DK East system demand with different EV penetration for Method 3.

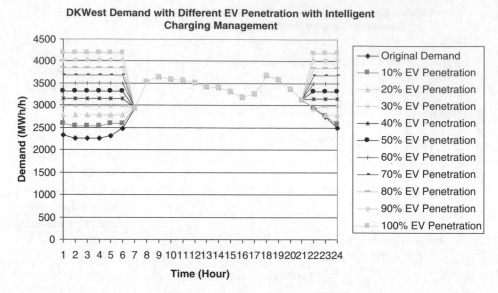

Figure 7.15   DK West system demand with EV charging for intelligent EV charging schedule.

7.3.2   *Intelligent Charging Schedule Based on Spot Price*

In order to reduce the new peak demand created by the EV charging and increase the EV penetration level without putting extra stress on the existing system, an intelligent charging schedule based on spot price is proposed. The idea is to smooth the demands for the time

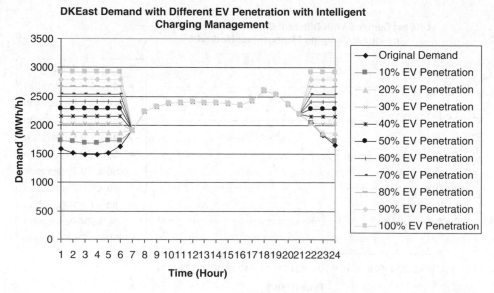

Figure 7.16   DK East system demand with EV charging for intelligent EV charging schedule.

periods when EVs are charging. The charging time periods are chosen as 00:00–06:00 and 22:00–24:00. The demands for these time periods will be same for the chosen charging time periods unless the resultant demand is less than the original demand.

The system demands with the proposed intelligent EV charging schedule method show that the EV penetration level can be increased to approximately 70% without putting extra stress on the existing system.

7.4 Analysis of Regulating Power from EVs

The analysis of regulating power from EVs is done to determine the available regulating power capacity from EVs based on the day-ahead EV charging schedule. The regulating power requirement and price will be used to calculate the economic return to EV users by providing regulating power.

7.4.1 Regulating Power Requirement and Price Analysis

The intention of the regulation power requirement and price analysis is to analyse the relation between wind power production and the regulating power requirement and the relation between the regulating power requirement and the regulating power prices.

Because the regulating power requirement data of DK East are not available from the Energinet.dk website and there is a big deviation between the data from Energinet.dk and the Nord Pool Spot website, it is decided to choose the regulating power requirement data of DK West from Energinet.dk to carry out the regulating power requirement and price analysis. Therefore, the wind power production data, the regulating power data and the regulating price data of DK West in 2010 were obtained from the Energinet.dk website.

One of the objectives of integrating a large number of EVs into power grids is to help integrate more renewable energy into power systems. In Denmark, wind power is the main renewable energy resource. The target of the Danish government is that approximately 50% electricity demand will be supplied by wind power and the wind power capacity will be doubled in 2020 [6]. The intention of analysing the relation between wind power production and regulating power requirements is to estimate what the regulating power requirements will be in 2020 with the 50% wind power target realized.

In order to get the relation between wind power production and the regulating power requirement, the plots of wind power production versus up-regulation and wind power production versus down-regulation were obtained and are shown in Figure 7.17 and Figure 7.18, respectively. These figures show that the relation between the wind power production and the regulating power requirements is not very obvious. But it is reasonable to assume that the regulating power requirement can be doubled with the doubled wind power capacity in the situation when the regulating power requirements are high due to the wind power production prognosis.

From the EV users' perspective, the objective of providing regulating power is to gain as much economic return as possible. Therefore, the relation between the regulating requirements and the regulating prices is analysed as well. The relations between up-regulation and up-regulation prices and between down-regulation and down-regulation prices are illustrated in Figure 7.19 and Figure 7.20, respectively. It is shown in the plots that the

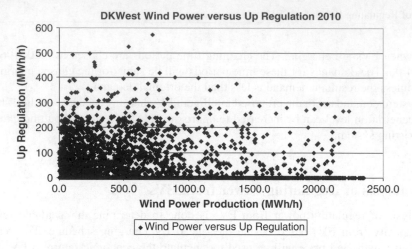

Figure 7.17 DK West wind power versus up-regulation in 2010.

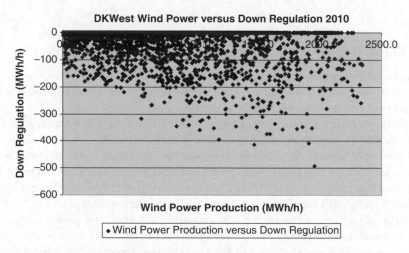

Figure 7.18 DK West wind power versus down-regulation in 2010.

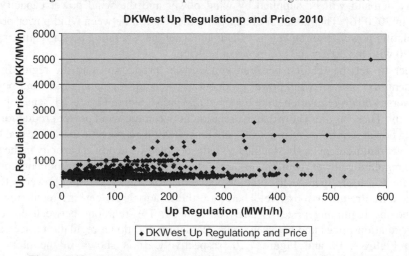

Figure 7.19 DK West up-regulation versus up-regulation price in 2010.

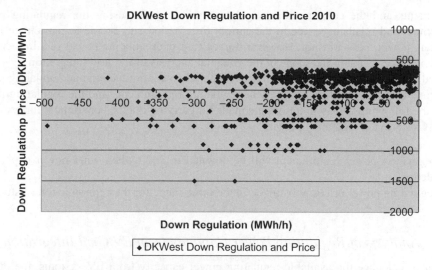

Figure 7.20   DK West down-regulation versus down-regulation price in 2010.

very high up-regulation prices and the very low down-regulation prices happened during high up-regulation requirement time periods and high down-regulation requirement time periods.

The DK West regulating power and regulating power prices are illustrated in Figure 7.21. This shows that up-regulation or down-regulation was needed in consecutive time periods. Therefore, for EV users and the EV fleet operator, they can forecast the regulating power

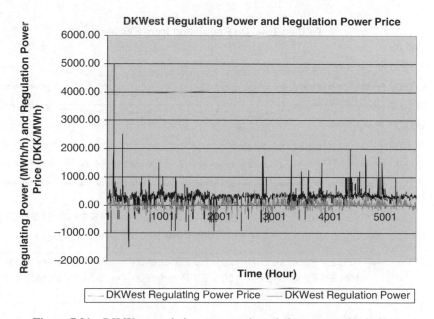

Figure 7.21   DK West regulating power and regulating power price in 2010.

requirements and the corresponding regulating power prices using the regulating power requirement and regulating price in the current time period and decide to participate in the regulating power market or not. However, forecasting the regulating power requirement and regulating prices is not in the scope of the potential analysis of EV grid integration.

Based on the analysis of the relation between wind power production and regulating power requirement, and the relation between the regulating requirements and the regulating power prices, the following two assumptions are made for carrying out the economic return study of having EVs provide regulating power:

- the regulating power requirement will be doubled in 2020 when wind power capacity is doubled;
- the regulating power prices will still be in the same range for the economic return study.

7.4.2   Analysis of Regulating Power Capacity from EV Grid Integration

In order to calculate the available regulating power capacity from EV systems, the driving pattern analysis results are used to get the available EVs in each hour of a day.

The regulating power consists of both up regulation and down regulation. The up regulation can be provided by releasing charging schedule, feeding power back to grid and both.

Based on the charging study in Section 7.3, the EVs which are doing charging at each hour are calculated for the intelligent charging scenario. This group of EVs can provide up-regulation by charging release. The EVs which do not have a charging schedule can provide up-regulation by feeding power back to the grid if needed. The total up-regulation capacity is the sum of the two up-regulations.

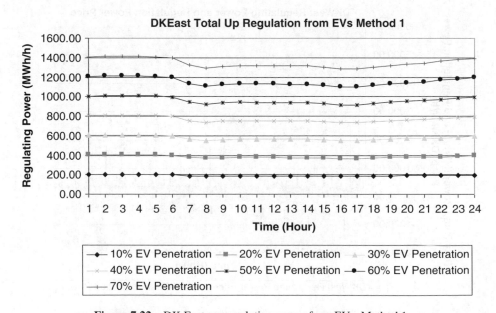

Figure 7.22   DK East up-regulating power from EVs, Method 1.

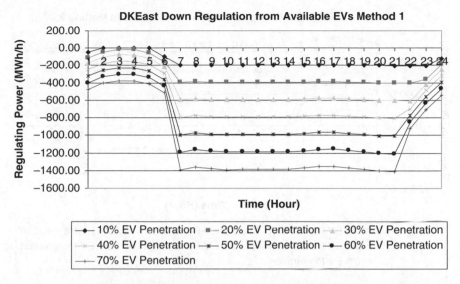

Figure 7.23 DK East down-regulating power, Method 1.

The down-regulation can be provided by charging the EVs which do not have a charging schedule in a specific hour.

Since DK West and DK East are connected to different synchronizing zones, the regulating power capacity analysis was done for DK West and DK East independently.

The available regulating power capacities in DK East from EVs for the three charging methods were obtained and are illustrated in Figures 7.22–7.27.

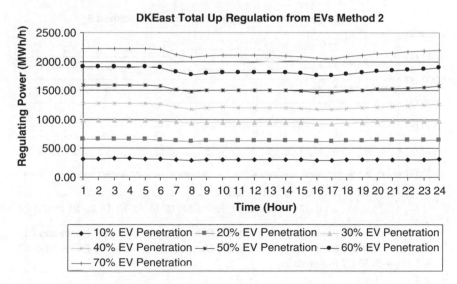

Figure 7.24 DK East up-regulating power from EVs, Method 2.

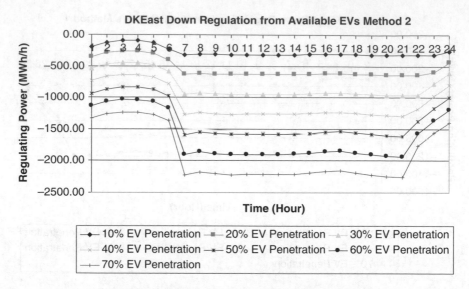

Figure 7.25   DK East down-regulating power, Method 2.

The maximum and minimum DK East up-regulating power and down-regulating power capacities with different EV penetration for Method 1, Method 2 and Method 3 are listed in Table 7.6 and Table 7.7, respectively.

The maximum up-regulation capacity in DK East is 6575.62 MWh/h with 70% EV penetration and Method 3, and the minimum up-regulation is 183.16 MWh/h with 10% EV penetration and Method 1.

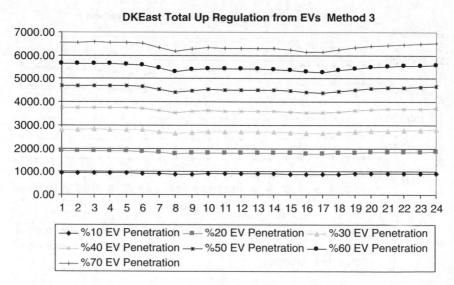

Figure 7.26   DK East up-regulating power from EVs, Method 3.

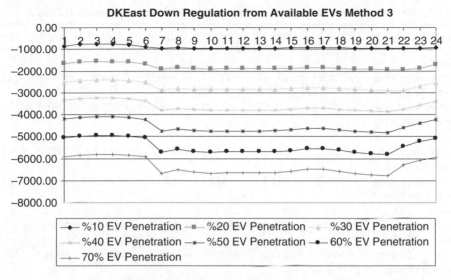

Figure 7.27   DK East down-regulating power, Method 3.

The maximum down-regulation capacity in DK East is −6775.25 MWh/h with 70% EV penetration and Method 3, and the minimum up-regulation is 0.00 MWh/h with 10% EV penetration and Method 1.

Since the driving pattern analysis was done for the whole country, the up-regulating power and down-regulating power from EVs, which is very much dependent on the driving pattern, are pretty much the same as for DK East.

Table 7.6   DK East maximum and minimum up-regulating power for Methods 1, 2 and 3.

EV penetration level	10%	20%	30%	40%	50%	60%	70%
Method 1							
Max up-regulating power (MWh/h)	204.35	407.9	609.26	809.94	1010.49	1211.05	1411.6
Min up-regulating power (MWh/h)	183.16	366.32	549.48	732.64	915.81	1098.97	1282.13
Method 2							
Max up-regulating power (MWh/h)	322.33	653.93	980.9	1275.87	1592.9	1909.94	2226.97
Min up-regulating power (MWh/h)	293.06	616.96	925.45	1172.23	1465.29	1758.35	2051.4
Method 3							
Max up-regulating power (MWh/h)	943.56	1883.34	2822.41	3760.81	4699.08	5637.35	6575.62
Min up-regulating power (MWh/h)	879.17	1758.35	2637.52	3516.69	4395.87	5275.04	6154.21

Table 7.7 DK East maximum and minimum down-regulating power for Methods 1, 2 and 3.

EV penetration level	10%	20%	30%	40%	50%	60%	70%
Method 1							
Max down-regulating power (MWh/h)	−201.99	−403.41	−605.11	−806.58	−1008.2	−1209.9	−1411.51
Min down-regulating power (MWh/h)	0	−16.07	−76.05	−149.47	−225.49	−301.5	−377.52
Method 2							
Max down-regulating power (MWh/h)	−323.18	−645.45	−968.17	−1290.5	−1613.2	−1935.8	−2258.42
Min down-regulating power (MWh/h)	−92.79	−261.3	−443.88	−639.92	−838.55	−1037.2	−1235.81
Method 3							
Max down-regulating power (MWh/h)	−969.53	−1936.4	−2904.5	−3871.6	−4839.5	−5807.4	−6775.25
Min down-regulating power (MWh/h)	−746.72	−1569.2	−2405.7	−3255.7	−4108.1	−4960.1	−5812.02

The maximum and minimum DK West up-regulating power and down-regulating power capacities with different EV penetration for Method 1, Method 2 and Method 3 are listed in Table 7.8 and Table 7.9, respectively.

The maximum up-regulation capacity in DK West is 8928.31 MWh/h with 70% EV penetration and Method 3, and the minimum up-regulation is 248.71 MWh/h with 10% EV penetration and Method 1.

Table 7.8 DK West maximum and minimum up-regulation for Methods 1, 2 and 3.

EV penetration level	10%	20%	30%	40%	50%	60%	70%
Method 1							
Max up-regulating power (MWh/h)	277.48	553.94	827.35	1099.4	1371.72	1644.04	1916.36
Min up-regulating power (MWh/h)	248.71	497.41	746.12	994.82	1243.53	1492.23	1740.94
Method 2							
Max up-regulating power (MWh/h)	437.53	870.27	1301.85	1732.05	2162.54	2593.02	3023.51
Min up-regulating power (MWh/h)	397.93	795.86	1193.78	1591.71	1989.64	2387.57	2785.5
Method 3							
Max up-regulating power (MWh/h)	1281.07	2557.36	3832.48	5106.23	6380.26	7654.29	8928.31
Min up-regulating power (MWh/h)	1193.79	2387.57	3581.36	4775.14	5968.93	7162.71	8356.5

Table 7.9 DK West maximum and minimum down-regulating power for Methods 1, 2 and 3

EV penetration level	10%	20%	30%	40%	50%	60%	70%
Method 1							
Max down-regulation (MWh/h)	−273.8	−547.76	−821.65	−1095.2	−1369	−1642.8	−1916.62
Min down-regulation (MWh/h)	0	−20.38	−101.82	−210.6	−313.82	−417.04	−520.26
Method 2							
Max down-regulation (MWh/h)	−438.83	−876.42	−1314.6	−1752.3	−2190.4	−2628.5	−3066.59
Min down-regulation (MWh/h)	−128.85	−353.36	−601.28	−876.56	−1146.3	−1416	−1685.68
Method 3							
Max down-regulation (MWh/h)	−1316.5	−2629.3	−3943.9	−5257	−6571.3	−7885.5	−9199.77
Min down-regulation (MWh/h)	−1016.8	−2129.2	−3265.1	−4428.3	−5586	−6743.6	−7901.26

The maximum down-regulation capacity in DK West is -9199.71 MWh/h with 70% EV penetration and Method 3, and the minimum up-regulation is 0.00 MWh/h with 10% EV penetration and Method 1.

The analysis of the regulating power capacity from EVs shows that the EVs are a big regulating power resource for the system operator to handle the bigger fluctuation that will be introduced by using more renewable energy. The limitation of the regulating power capacity analysis for EV grid integration is that the power system network constraints are not included. This should be taken into account in future work.

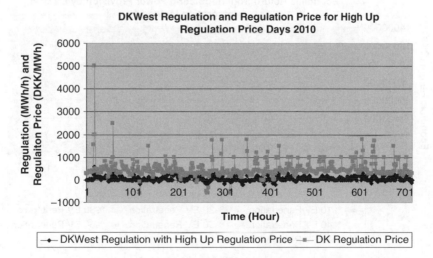

Figure 7.28 DK West days with very high up-regulation prices in 2010.

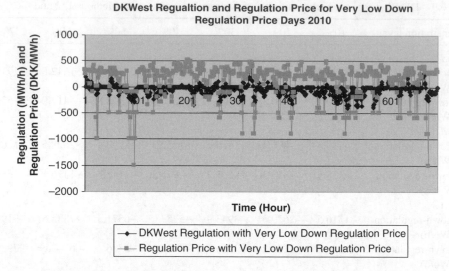

Figure 7.29   DK West days with very low down-regulation prices in 2010.

7.4.3   Economic Return from Regulating Power Provision by EVs

In order to investigate the economic return for EV users by providing regulating power, the days with very high up-regulation prices and very low down-regulation prices in 2010 of DK West were selected from the extracted regulating power data from Energinet.dk. The regulating power and regulating power prices for the selected days are illustrated in Figure 7.28 and Figure 7.29, respectively.

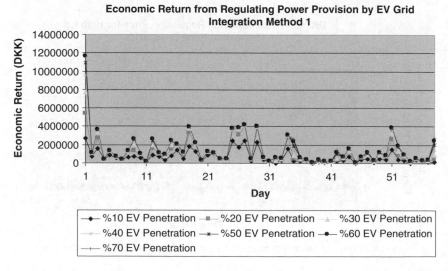

Figure 7.30   Economic return to EV users by providing regulating power, Method 1.

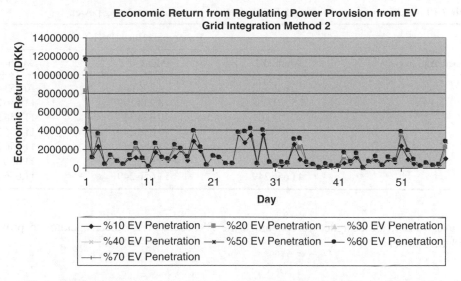

Figure 7.31   Economic return to EV users by providing regulating power, Method 2.

The economic return of regulating power provision by EVs was calculated for the three charging methods with different EV penetrations. The results are illustrated in Figure 7.30, Figure 7.31 and Figure 7.32. The maximum economic return with different EV penetration for the three charging methods is listed in Table 7.10.

The economic return to the EV users by providing regulating power is not very high considering the EV numbers. Therefore, a new attractive economic incentive scheme for

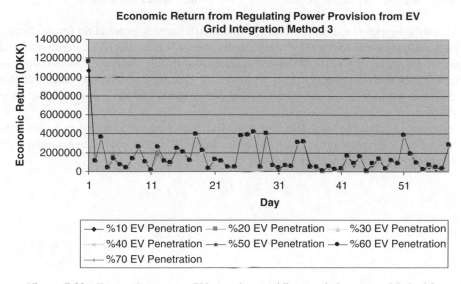

Figure 7.32   Economic return to EV users by providing regulating power, Method 3.

Table 7.10   Maximum economic return for EV users by providing regulating power.

EV penetration (%)	Economic return (DKK)		
	Method 1	Method 2	Method 3
10	2 685 532	4 296 844	10 653 449
20	5 371 068	8 234 199	11 656 249
30	7 774 109	10 653 449	11 656 249
40	9 630 149	11 656 249	11 656 249
50	10 909 249	11 656 249	11 656 249
60	11 656 249	11 656 249	11 656 249
70	11 656 249	11 656 249	11 656 249

EV users has to be designed to encourage the EV users or EV fleet operator to provide regulating power.

7.5   Summary

The purpose of the analysis of regulating power from EVs is to investigate the possible regulating power capacity from EV systems and the economic return to EV users by providing regulating power.

In order to achieve the study objective of the potential study of EV grid integration, the following work has been done and is presented in this chapter:

- driving pattern analysis for EV grid integration;
- EV charging schedule based on spot price;
- potential analysis of regulating power provision from EV grid integration.

The regulating power capacities from EVs with different EV penetration for the three EV charging methods used were studied and it can be concluded that the EVs can provide a large amount of regulating power for the system operators to handle the bigger fluctuation caused by higher renewable energy penetration. However, the economic return study results show that the economic return with the current regulating power market set-up is not attractive enough to EV users.

The power system network constraints have to be included in future work to get a more accurate value of the regulating power capacities from EVs and the relevant economic return to EV users by providing regulating power.

References

[1] Brassin, J. (2007) Vehicle-to-grid improving wind power integration, Master thesis, Center for Electric Technology, DTU.

[2] Chandrashekhara, D.K., Østergaard, J., Larsen, E. *et al.* (2010) Integration of electric drive vehicles in the Danish electricity network with high wind power penetration. *European Transactions on Electric Power*, **20** (7), 872–883.

[3] Brooks, A.N. (2002) Vehicle-to-grid demonstration project: grid regulation ancillary service with a battery electric vehicle. Final report, AC Propulsion, Inc.

[4] Kempton, W. and Tomic, J. (2005) Vehicle-to-grid fundamentals: calculating capacity and net revenue. *Journal of Power Sources*, **144** (1), 268–279.

[5] Andersson, S.L., Elofsson, A.K., Galus, M.D. *et al.* (2010) Plug-in hybrid electric vehicles as regulating power providers: case studies of Sweden and Germany. *Energy Policy*, **38** (6), 2751–2762.

[6] EA Energy Analyses, 50% Wind power in Denmark in 2025, http://www.ea-energianalyse.dk/projects-english/642_50_per_cent_wind_power_in_Denmark.html (accessed January 2013).

8

Frequency-Control Reserves and Voltage Support from Electric Vehicles

Jayakrishnan R. Pillai and Birgitte Bak-Jensen
Department of Energy Technology, Aalborg University, Aalborg, Denmark

8.1   Introduction

Electric power ancillary services are essential for the stable and reliable operation of power systems. The transmission system operator (TSO) uses a series of ancillary services in order to continuously balance the aggregated generation and demand, to manage transmission line flows and to ensure acceptable voltages in the power system. These support functions are activated over different time frames which are necessary during normal, contingency and restoration states of the power system operation. The ancillary services are an integral part of the service provided by the vertical-integrated utilities. However, the introduction of the competitive and deregulated electricity market has provided clear definition, quantification and dedicated markets for these services. This creates additional revenues to the generator owners and is an efficient way of optimizing the system reliability cost. To ensure the security of the power system operation, it is the TSO who is mainly responsible for the purchase and management of these services, which are acquired mostly from the conventional power plants. With the introduction of smart grids, the information and communication technology (ICT), aggregators and flexible market enables the loads to be a potential and active supplier of most of the ancillary services.

This chapter describes the role of electric vehicles in providing active power balancing support in power system operation. The ancillary services that are relevant for the electric vehicles are introduced. Case studies employing electric vehicles for active power balancing in the typical Danish power systems, characterized by large amounts of wind power are discussed.

Grid Integration of Electric Vehicles in Open Electricity Markets, First Edition. Edited by Qiuwei Wu.
© 2013 John Wiley & Sons, Ltd. Published 2013 by John Wiley & Sons, Ltd.

The other supplementary grid support functions from electric vehicles, like voltage control, are briefly presented. The various concepts discussed in this chapter are mostly based on the Danish power system context, which reasonably represents the model for future electricity grids with higher penetration of wind turbine and other dispersed generation units.

8.2   Power System Ancillary Services

Many and varied definitions exists for ancillary services. Taking into account the changing requirements of the electricity markets in Europe, with increasing penetration of renewable energy sources, Eurelectric [1] gives the following definition for ancillary services:

> Ancillary services are all services required by the transmission or distribution system operator to enable them to maintain the integrity and stability of the transmission or distribution system as well as the power quality.

In general, the ancillary services deal with monitoring, automatic or manual operation and control of resources for the reliable and quality operation of power system. The typical way of receiving ancillary services is by a combination of administrative tools like grid codes which are enforced and market solutions like hourly trade, short- and long-term contracts and so on [2]. There are many grid services defined and used as ancillary services in different countries. The seven important ancillary services considered by Eurelectric are frequency control, voltage control, spinning reserve, standing reserve, black start capacity, remote automatic generation control and emergency control action. The common definitions for these services are necessary for cross-border trading. The frequency control, voltage control and emergency control action for each dispatchable unit are chosen mandatory to ensure the quality of power supply [1].

The first four ancillary services listed by Eurelectric are considered mandatory in Denmark. The frequency and voltage control are essential for the normal operation of the system, whereas the spinning and standing reserves are used for the failure prevention of the system. The active power services are supplied by the centralized and decentralized conventional generators and the strong interconnections to the neighbouring countries. The voltage control is realized by the reactive power compensation (static Var compensators, under/over-excitation of generators) in the local networks. The TSO in Denmark, Energinet.dk, procures these services which are provided mainly by the generators and the network devices (for voltage control) [3]. These services are procured under negotiated contracts or electricity market and their pricing is regulated and monitored by the Danish Energy Regulatory Authority. The price for the ancillary services is recovered from the customers and generators as system tariffs [1].

8.3   Electric Vehicles to Support Wind Power Integration

The increasing penetration of renewable energy units into the power system, especially the wind turbines, which are matured and proven technology, has a significant influence on the changing structure of the ancillary services market. The wind power is highly variable and unpredictable and, thus, has poor load-following characteristics. This results in increased need for balancing power. The modern wind turbines are equipped with suitable controls to provide the grid support functions. However, these provisions are not commonly used owing to the fact that the 'clean' and 'free' wind energy is spilled during curtailment. Moreover, the conventional generators, which are the common sources that supply the ancillary services, are gradually phased out by the increasing wind power integration. In Denmark, the wind power supplies more than 20% of the average annual consumption [4]. Denmark plans to increase the wind power production to 50% by 2020, which in turn reduces the capacity of large centralized power plants by 40% [5, 6]. Further, the capacity expansion of the interconnectors with the neighbouring areas may be limited, as large wind turbine installations are also planned across the borders. So it is obvious that the increment in the share of wind power is a function of the ability of the power system to regulate the supply. These factors contribute to the need for alternate, fast-acting and local technologies to deliver the increasing volumes of ancillary services in power systems.

Battery storage is one of the best alternate solutions to compensate the intermittency of wind power generation. The batteries could store surplus electricity in the grid and deliver back to the grid on deficit of generation. This property of battery storage is complementary to the variable nature of wind power and could be used to smooth both short-term and long-term wind power variations. The battery storage could also provide power quality control functions and other major utility ancillary services, like frequency control or active power balancing. The use of the battery storage of electric vehicles is one of the widely discussed and evolving concepts that could act as a load reacting to changes in the power supply. A significant fleet of electric vehicles when aggregated by a 'fleet operator' or 'aggregator', with the support of local intelligence and communication, can provide temporary distributed electricity storage in the power system when they are not used for driving [7]. The electric-vehicle batteries could charge as a controllable load to store energy during high winds and low electricity prices and could also discharge as quick-response generation when required. The reliability of the renewable electricity will be enhanced with the vast untapped distributed storage of electric vehicle fleets when connected to the grid. This could be considered as a large aggregated megawatt battery storage which is collectively termed as 'vehicle-to-grid' (V2G) system [8, 9]. The TSO could request for power transfer to facilitate ancillary services like automatic generation control or frequency control and operating reserves from the fleet of vehicle batteries through the aggregators.

In Denmark, currently there are instances where the wind power generation exceeds the total consumption. With 50% wind penetration in the future power system, this is expected to cross 1000 h in a year. More than half of the power imbalances in Denmark are caused by wind power, where approximately 70% are caused by wind prediction errors [10]. A wind forecast error of 1 m s^{-1} results in a prognosis error of 450 MW active power production, which is quite significant comparing the size of the Danish power system [11]. The integration of more wind turbines in the Danish power system to meet the 2020 Danish Energy plan introduces more stochastic and variable electricity production. Additional reserve power capacity is needed to fill in the variance between predicted and scheduled wind generation. This creates an ideal market situation for utilizing the ancillary services from electric-vehicle batteries in the Danish power system. Studies conducted for the US and Danish electricity markets indicate that electric vehicles are ideally suited and economically viable to provide ancillary services like active power balancing reserves [9, 12]. The renewable energy initiatives, active promotion of smart grids and local distributed energy resources are the major driving forces for the electric vehicle market in Denmark [13].

8.4   Electric Vehicles as Frequency-Control Reserves

The system frequency is the primary indicator of the power balance in a power system. The frequency deviations in a power system reflect the difference between the generation and consumption. This mismatch is caused by the difference in the scheduled and the actual power and the disturbances in the system. The frequency-control reserves are operated to deal with these deviations and disturbances. The West Denmark (DK1) power system is part of the Regional Group (RG) Continental Europe and East Denmark (DK2) is part of the Regional Group Nordic. Table 8.1 gives the different frequency-control reserves in DK1 and DK2 respectively [14]. The frequency-control schemes are generally classified as primary, secondary and tertiary reserves, as practised in the DK1 power system. The identification, design and activation of these reserves differ from country to country or regions due to the different levels of electricity market development.

The primary reserves are used as an instantaneous frequency reserve to deal with sudden power imbalances. The secondary reserves operate a bit slower than the primary reserves, which help to restore the nominal system frequency and minimize any power exchange deviations between control areas. The manual or tertiary reserves are the slowest of all the control reserves and are used to restore the secondary reserves by rescheduling the generation. Figure 8.1 shows the frequency-control schemes and processes followed in the RG Continental Europe electricity system.

Table 8.1   Frequency control reserves in DK1 and DK2 for the year 2011.

Reserve types	Capacity (MW)	Activation period	Activation mode and purchase
DK1 as part of RG Continental Europe			
Primary reserves	±27	0–30 s	Automatic: droop control (2–8%), activated at 49.8–50.2 Hz, daily auctions
Secondary reserves	±90	30 s–15 min	Automatic: load frequency control (LFC), monthly contracts
Tertiary reserves	~450	15 min	Manual, daily auctions
DK2 as part of RG Nordic			
Frequency-controlled normal operating reserves	±23	2–3 min	Automatic, activated at 49.9–50.1 Hz, daily auctions
Frequency-controlled disturbance reserves	±150	50% in 5 s and remaining 50% in the next 25 s	Automatic, activated at 49.9–49.5 Hz, daily auctions
Manual reserves	~600	15 min	Manual, daily auctions

Source: Energinet.dk [14].

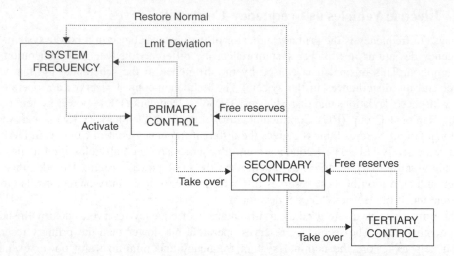

Figure 8.1 Frequency control scheme of RG Continental Europe. (*Source:* ENTSOE [15].)

The capacity reserves of generation units are commonly used for frequency-control services. Their prices are quite a bit higher than energy prices and are volatile due to opportunity costs. By participating in the frequency-control reserve market, electric-vehicle owners could gain higher profits than from the energy market. Of all the reserves, the automatic reserves are the most expensive as they receive both reserve and activation payments and, therefore, are best suited for electric vehicles. With more wind turbines being introduced into the power systems, the limited predictability and higher fluctuations of wind power call for increasing amounts of frequency reserves. In this context, local resources like electric vehicles with faster ramp rates, quick start and flexible characteristics are very relevant for the reliable operation of the power system. In order to accommodate these smaller and local units like electric vehicles, potential changes are expected in the framework and operation of electricity markets for balancing reserves. These changes include shorter time frames for reserve activation and moving the price settings of reserve power closer to the operating period [16].

8.4.1 Primary Reserves

The primary reserves are typically supplied by the governors of the power plant units within the power system control area. The power plant output is changed in proportion to the frequency changes:

$$\Delta P_{\mathrm{G}} = -\frac{1}{K}\Delta f \tag{8.1}$$

where ΔP_{G} (MW) is the generation change, K (%) is the governor droop and Δf (Hz) is the frequency change.

The droop characteristics of the governors are adjusted to a new operating point by which the frequency deviations are minimized. The amount of primary reserves in large interconnected power systems is quite low, as shown in Table 8.1, where the capacity reserves from the

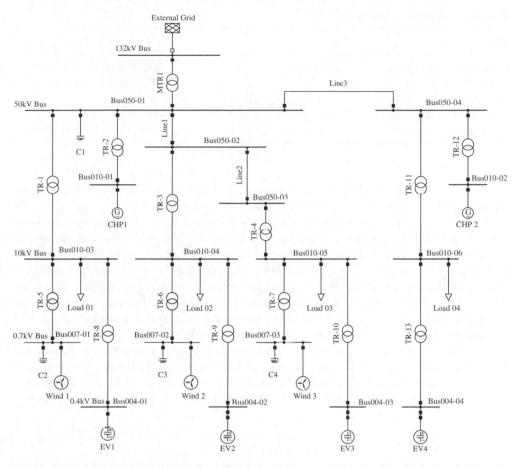

Figure 8.2 Test distribution network. (Reproduced from [18], J.R. Pillai and B. Bak-Jensen (2010) 'Vehicle-to-grid systems for frequency regulation in an islanded Danish distribution network', in *2010 IEEE Vehicle Power and Propulsion Conference (VPPC)*, by permission of IEEE.)

conventional generators may be adequate to supply this grid service. The primary reserves are fully activated within very short periods, but this time frame of operation may be irrelevant for large power imbalances resulting from wind power fluctuations in the grid. Utilizing electric-vehicle batteries for power balancing in such situations may turn uneconomical and complex with respect to reserve activation, communication and coordination. However, in smaller networks like islanded systems, isolated distribution grids or sub-grid networks (as demonstrated by the cell controller project [17]), the role of electric vehicles could be more prominent as primary reserves. To illustrate this, the relevant findings of a case study simulated for a characteristic Danish distribution system are presented here. The aggregated electric vehicle storages perform the primary frequency-control function by acting either as a charging load or generator. Figure 8.2 depicts a simplified grid model of a medium-voltage Danish distribution network, islanded from the 132 kV external grid. The model is implemented in DIgSILENT Power Factory software [19]. Table 8.2 gives the relevant details of the grid components.

Table 8.2   Distribution grid details.

Units	Type	Total capacity and control mode
Combined heat and power (CHP) – two units	Gas turbine	CHP1: 20 MW (isochronous) CHP2: 4 MW (droop control)
Wind farms – three units	Fixed-speed turbines	11.5 MW
Aggregated electric vehicle storages – four units	Static generator (ideal PWM converter model) [19]	4 MW (droop control), initial charging load (0.4 MW/unit)

The total connected load on the distribution grid is 24 MW, where 6.6% constitutes the electric-vehicle charging load. The distribution network is a wind-power-dominated grid, with 48% of the total load supplied by the wind turbines. The largest CHP unit operates in an isochronous mode [20], where the machine will control the governor to maintain the system frequency, irrespective of the load. The other CHP unit operates in droop mode, where the control gain is the inverse of the governor droop. The aggregated electric-vehicle storages are also operated in droop mode to adjust the active power levels to respond to the system frequency deviation. The battery charges for positive frequency deviations and discharges (or reduces the charging levels) as the grid frequency drops. Figure 8.3 shows the system frequency response of the distribution network for a simulation case, where a step load change of 2 MW is applied at $t = 5$ s. The rate of change of frequency and the minimum frequency drop are significantly less for the simulation case with electric vehicles participating in primary frequency control. The electric-vehicle storage units have only very small delays when compared with the dynamics of a conventional generation unit, which gives the former a more active role in frequency control.

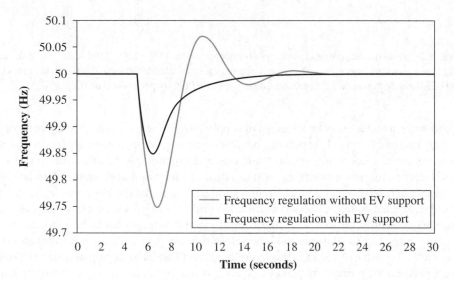

Figure 8.3   System frequency response for a step load change (Reproduced from [18], J.R. Pillai and B. Bak-Jensen (2010) 'Vehicle-to-grid systems for frequency regulation in an islanded Danish distribution network', in *2010 IEEE Vehicle Power and Propulsion Conference (VPPC)*, by permission of IEEE.)

The electric-vehicle battery thus provides a more stable, better damped and fast recovery of the system frequency. To fulfil the primary frequency support, the aggregated electric-vehicle storage has reduced its initial charging from 1.6 MW to 0.52 MW, thus acting as a controllable load. A further larger frequency excursion enables the storages to discharge power back to the grid, realizing the V2G function.

8.4.2   Secondary Reserves

The secondary reserves generally use centralized and automatic frequency control to match the power balance between the generation and demand and to maintain the scheduled power exchanges between the interconnectors. The controller adjusts the active power set points of the generation or consumption resources participating in this process. The procedure is commonly referred to as automatic generation control or load frequency control (LFC). To analyse the participation of electric-vehicle batteries as secondary reserves, the LFC in West Denmark (DK1) is considered here. The western Danish control area is connected to the large RG Continental European grid through the interconnector with northern Germany. The LFC in DK1 has to ensure that the power exchange deviations between these two control areas are kept to a maximum of ±50 MW [21, 22].

The western Danish power system is characterized by a large amount of dispersed generation units connected to 60 kV or below; this includes mostly gas-turbine-based CHP and wind turbine units. Out of the total generation capacity of approximately 8 GW, about 22% are CHP units and 35% are wind turbines (5% are offshore wind). The large coal-fired power plants constitute the remaining generation capacity. DK1 is also interconnected with RG Nordic through HVDC connections with Sweden, Norway and East Denmark. Figure 8.4 shows the block diagram of a load frequency controller integrating V2G, which is represented as aggregated electric-vehicle storage. The storage capacity is considered as 90 MW, 360 MWh, which is equivalent to 9000 electric vehicles (average power connection rating of 10 kW). This corresponds to less than 1% of the total fleet of 1 million cars in West Denmark. The generation, demand and interconnectors are modelled as aggregated units in the LFC [24]. The power imbalance in an interconnected system is reflected as area control error (ACE):

$$\text{ACE} = (P_{\text{meas}} - P_{\text{sch}}) + B(f_{\text{meas}} - f_0) \tag{8.2}$$

$$\therefore \text{ACE} = \Delta P_{\text{t}} + B\Delta f \tag{8.3}$$

where B is the frequency bias factor (depends on the load sensitivities and governor response characteristics of the control area), P_{meas} is the measured power exchange, P_{sch} is the scheduled power exchange, Δf is the frequency deviation ($f_{\text{meas}} - f_0$), and ΔP_{t} is the total power deviation between the interconnected control area.

The load frequency controller integrates the control error to generate the ramp loading of the units (ΔP_{ref}), which is the average power distributed among the units participating in the secondary frequency control:

$$\Delta P_{\text{ref}} = -\text{ACE} \cdot \beta - \frac{1}{T} \int \text{ACE} \, dt \tag{8.4}$$

where β is the the proportional gain of the controller and T is the controller time constant.

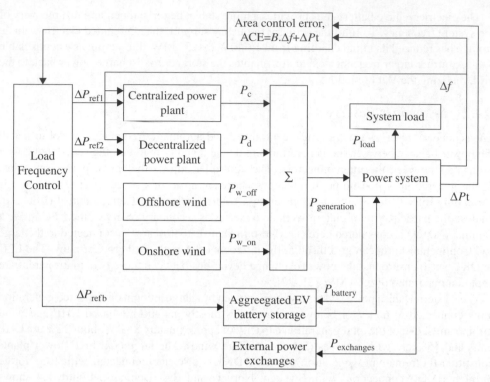

Figure 8.4 Block diagram of load frequency controller including aggregated electric vehicle storage. (Reproduced from [23], J.R. Pillai and B. Bak-Jensen (2011) 'Integration of vehicle-to-grid in the western Danish power system', *IEEE Transactions on Sustainable Energy*, **2** (1), 12–19, by permission of IEEE.)

The V2G systems are quite free from ramp limitations when compared with conventional generators [24]. Therefore, battery storage in LFC has fast power regulation characteristics compared with conventional generators. However, the reserve capacity of the V2G systems may be limited by state of charge of the storage. These aspects are analysed here, where LFC order (ramp loading) of the conventional generators is determined from the insufficiency of the V2G in meeting the power regulation requirement. Figure 8.5 shows the LFC simulation results of power exchange deviation between West Denmark and North German control areas for a typical windy winter day, where the wind power supplies an average 40% of the total demand in DK1. The day was also characterized by the total generation exceeding the consumption and more instants of down-regulation LFC signals.

A positive power exchange deviation in Figure 8.5 indicates less planned power being exchanged and the negative value gives the surplus power exchanged with Germany. The case without the V2G support shows large power exchange deviations which exceed the desired levels of ±50 MW. The LFC simulation case with V2G support shows that the deviations are largely reduced well within the accepted levels. Towards the end of the day, the power regulation capability is lost for few hours, when the battery is full, which demands for a higher storage capacity. However, it is evident that the electric-vehicle storage can respond and act faster than conventional generators, while providing secondary reserves. The power exchange

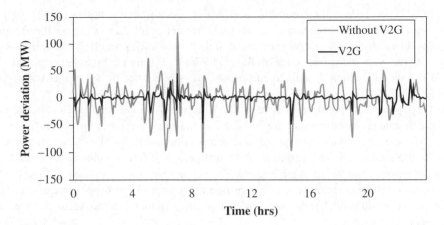

Figure 8.5 Power exchange deviations between West Denmark and Germany. (Reproduced from [23], J.R. Pillai and B. Bak-Jensen (2011) 'Integration of vehicle-to-grid in the western Danish power system', *IEEE Transactions on Sustainable Energy*, **2** (1), 12–19, by permission of IEEE.)

deviations are highly minimized with the V2G operating as secondary reserves, which is relevant for the reliable and economical operation of the control areas.

The simulation and the demonstration results of the Danish project 'Electricity demand as frequency controlled reserve (DFR)', described in Section 3.3.6, is an ideal step forward for encouraging the implementation of electric vehicles to provide active power reserves in the Danish power systems [25]. The project considers the use of thermostatically controlled heaters and refrigerators to function as frequency controlled reserves by reducing their consumption. Figure 8.6 gives the simulation results of frequency response of the Nordic power system during an outage of a 500 MW generator at $t = 1500$ s [26]. The electric heaters of total 1000 MW capacity are used here to function as frequency-controlled disturbance reserves

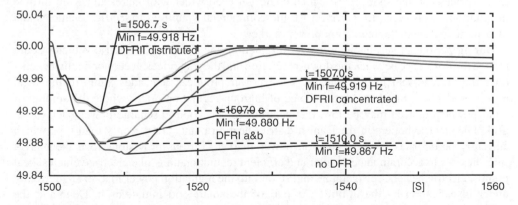

Figure 8.6 Frequency response of Nordic system under loss of generation applied with different DFR control logic. (Reproduced from [26], Z. Xu, J. Østergaard and M. Togeby (2011) 'Demand as frequency controlled reserve', *IEEE Transactions on Power Systems*, **26**, 1062–1071, by permission of IEEE.)

(49.9–49.5 Hz) to reduce their consumption during system frequency drops. The DFR II model, which continuously regulates its temperature set point depending on the frequency deviations, has reduced consumption the most in this case study, providing low-frequency dip and smooth recovery of the frequency profile [27]. The DFR II model gives better performance than DFR I models which deactivate and reconnect only at specific set points with specific delays [27].

The DFR project demonstration is carried out on the Danish island of Bornholm, where automatic frequency-control units are installed at the premises of approximately 200 customers. Automatic disconnection or reduced consumption of units like electric heaters, heat pumps, bottle coolers and refrigerators for a short period of few minutes is activated on frequency deviations from the acceptable range. The preliminary results published from the DFR demonstration set up in Bornholm, where refrigerators are used to supply frequency-control normal reserves (49.9–50.1 Hz), has provided positive demand response to the frequency deviations [28].

8.4.3   Tertiary Reserves

The tertiary reserves are generally activated manually by the TSO to deal with larger power imbalances, where the secondary reserve capacity is insufficient. This is performed by changing the working points of the participating generators and loads, and rescheduling the power exchange plans between interconnectors to free up the secondary reserves. In the Danish power system, more than 80% of the power imbalances are handled by the manual or tertiary reserves [29]. The automatic reserve capacity utilization is limited, which allows the preserved volume to be used during unexpected system disturbances and rapid power fluctuations from renewable energy units like offshore wind farms [30]. The manual reserves are commonly used to deal with the larger fluctuations in the power system that occur within the operating hour. As Denmark is a part of the Nordic electricity market, these reserves are traded as regulating power in the common Nordic regulating power market. The presence of high wind power in the Danish power system has generally produced large variations between the volatile regulation power prices and the spot prices, which could get even wider with increasing wind turbine installations. This scenario is perfect for the commercial viability of electrical vehicles and their participation in the regulation power market.

During periods of surplus wind power production, the charging of electric vehicles could reduce the difference between the down-regulation and the spot prices. It economically benefits both the electric-vehicle user and the wind power producer, and adds value to the wind power production. Similarly, during hours of high up-regulation prices where the amount of wind power produced may be low, the electric vehicles could discharge power back to the grid. The electric boilers in the Danish district heating plants are the only load type which currently participates in the regulation power market. Various provisions to make market rules flexible to accommodate smaller and efficient regulation units like electric vehicles, heat pumps, electric heaters, refrigerators and so on into the regulation power market are currently under investigation by the utilities. As part of the smart grid initiatives in Denmark, the implementation and experiences gained from the pilot projects and standardization, facilitate the continuous development of the market rules and regulation to accommodate these alternate balancing power regulation solutions [13].

8.5   Voltage Support and Electric Vehicle Integration Trends in Power Systems

Voltage support involves the reactive power control which is used to maintain the voltage levels within acceptable limits under all operating conditions throughout the power system. The reactive power is location specific, as the number of generation units varies within an area. So there exist no competitive markets for trading the voltage control and these are procured through long-term contracts. Transformer tap changers, capacitors, static-Var compensators and so on are the common network devices used by the utilities for voltage control and reactive power support. Some of the large central generation units and the network devices are currently utilized in Denmark to supply reactive reserves. The electric-vehicle battery inverter could operate as a static-Var compensator with appropriate four-quadrant operation that can provide voltage and dynamic reactive power support in the local grids. Some well-defined methods to provide voltage support by reactive power from solar photovoltaic inverters are practised based on different inverter set points of power factor, voltage or reactive power characteristics [31, 32]. These methods may be relevant for providing voltage support from electric-vehicle inverters. In the future electricity markets, payments for supporting voltage control reserves may be based on locational prices or capacity payments. It may also include opportunity costs to compensate for reducing real power regulation while increasing the reactive power support [33].

The most favourable trend in the promotion and adoption of electric vehicles is envisaged as a gradual progression from not overloading the grid by 'using timer control' to 'using price signal' to 'controlled load using demand signal' to 'distributed storage' [34]. The electric vehicles acting as a controllable load (unidirectional battery charger) to supply regulation power could be the first generation of V2G technology to be in place by the next 5 years. The wear and tear resulting from power cycling of the vehicle battery, providing bidirectional power transfer, is not well understood. To quantify the impact of these battery degradation costs on the overall cost–benefits from rendering power regulation needs to be investigated in detail. This factor could prompt the bidirectional V2G concept to be considered as a second-generation application of grid-connected vehicles. The technical norms for anti-islanding, power quality, protection and so on need to be established, for the grid-connection of electric vehicles in supplying power back to the grid. However, standards already exist for grid integration and coordination of distributed storages, such as IEEE 1547.3-2007, EN 50438, IEC 61850-7-420. Other standards, like VDE-AR-N 4105, IEEE Std. 929-2000, IEEE SCC21 and so on, clearly define practices and procedures for grid integration of distributed energy resources like solar PV, wind turbines and so on and could be closely applied for electric-vehicle batteries also. The other grid-support applications from electric vehicles like the voltage control, micro-grid support, low-voltage ride through, emergency power and so on could be utilized in the local networks, once the V2G concept is technically matured and economically viable.

8.6   Summary

This chapter emphasized the common ancillary services that are applicable to electric vehicles in supporting more wind power in power systems. The various frequency-control reserves are described, and the importance of electric vehicles in addressing the grid balancing issues for an increasing wind power penetration are discussed. Power system simulation studies based on short-term frequency excursions, LFC and demand as frequency control reserves are presented. The simulation results presented are used to analyse the importance of electric vehicles as an alternative to conventional generator reserves in the power system.

Acknowledgements

Some of the work presented in this chapter is part of the research projects 'Coherent energy and environmental system analysis (CEESA)', supported in part by The Danish Council for Strategic Research, and 'Tomorrow's high-efficiency electric car integrated with the power supply system', supported by The European Regional Development Fund (grant number, ERDFK-08-0011).

References

[1] Thermal Working Group (2004) Ancillary services: unbundling electricity products – an emerging market, 2003-150-0007, Eurelectric, February.
[2] Kirschen, D.S. and Strbac, G. (2004) *Fundamentals of Power System Economics*, John Wiley & Sons.
[3] Akhmatov, V. and Eriksen, P.B. (2007) A large wind power system in almost island operation – a Danish case study. *IEEE Transactions on Power Systems*, **22** (3), 937–943.
[4] WWEA (2010). World Wind Energy Report 2009, World Wind Energy Association, Bonn, http://www.wwindea.org/home/images/stories/worldwindenergyreport2009_s.pdf (accessed January 2013).
[5] Energinet.dk (2010) System Plan 2010, October.
[6] Energinet.dk (2007) System Plan 2007, October.
[7] Kempton, W. and Tomic, J. (2005) Vehicle-to-grid power fundamentals: calculating capacity and net revenue. *Journal of Power Sources*, **144** (1), 268–279.
[8] Kempton, W. and Tomic, J. (2005) Vehicle-to-grid power implementation: from stabilizing the grid to supporting large-scale renewable energy. *Journal of Power Sources*, **144** (1), 280–294.
[9] Kempton, W. and Tomic, J. (2007) Using fleets of electric-drive vehicles for grid support. *Journal of Power Sources*, **168** (1), 459–468.
[10] Agersbæk, G. (2007) Correlation between wind power and local CHP, http://www.giz.de/Themen/de/dokumente/en-dialog-agersbaek-2007.pdf (accesed January 2013).
[11] Agersbæk, G. (2010) Integration of wind power in the Danish energy system, http://www.nwcouncil.org/energy/wind/meetings/2010/07/WIF%20TWG%20072910%20Agersbaek%20072010.pdf (accesed January 2013).
[12] Larsen, E., Divya, K.C. and Østergård, J. (2008) Electric vehicles for improved operation of power systems with high wind power penetration, in *IEEE Energy 2030 Conference, 2008. ENERGY 2008*, IEEE, Piscataway, NJ.
[13] Energinet.dk (2010) Smart grid in Denmark, http://energinet.dk/SiteCollectionDocuments/Engelske%20dokumenter/Forskning/Smart%20Grid%20in%20Denmark.pdf (accessed February 2013).
[14] Energinet.dk (2010). Ancillary services for delivery in Denmark (in Danish), December, http://energinet.dk/DA/El/Saadan-driver-vi-elsystemet/Systemydelser-for-el/Sider/Systemydelserforel.aspx (accessed February 2013).
[15] European Network of Transmission System Operators for Electricity (2009) Load-Frequency control and performance, https://www.entsoe.eu/fileadmin/user_upload/_library/publications/entsoe/Operation_Handbook/Policy_1_final.pdf (accessed February 2013).
[16] Hay, C., Togeby, M., Bang, N.C. *et al.* (2010) Introducing electric vehicle into the current electricity market, Edison Deliverable 2.3, 25 May, http://www.edison-net.dk/Dissemination/Reports/Report_004.aspx (accessed January 2013).
[17] Lund, P. (2007) The Danish Cell Project – Part 1: background and general approach, in *IEEE Power Engineering Society General Meeting, 2007*, IEEE, Piscataway, NJ.

[18] Pillai, J.R. and Bak-Jensen, B. (2010) Vehicle-to-grid systems for frequency regulation in an islanded Danish distribution network, in *2010 IEEE Vehicle Power and Propulsion Conference (VPPC)*, IEEE, Piscataway, NJ.

[19] DIgSILENT (2009) Power Factory User Manual Version 14.

[20] Kundur, P. (1994) *Power System Stability and Control*, McGraw-Hill, New York, NY.

[21] Akhmatov, V., Rasmussen, C., Eriksen, P.B. and Pedersen, J. (2006) Technical aspects of status and expected future trends for wind power in Denmark. *Wind Energy*, **10**, 31–49.

[22] Akhmatov, V., Kjærgaard, J.P. and Abildgaard, H. (2004) Announcement of the large offshore wind farm Horns Rev B and experience from prior projects in Denmark, in Proceedings of European Wind Energy Conference (EWEC 2004), London.

[23] Pillai, J.R. and Bak-Jensen, B. (2011) Integration of vehicle-to-grid in the western Danish power system. *IEEE Transactions on Sustainable Energy*, **2** (1), 12–19.

[24] Brooks, A.N. (2002) Vehicle-to-grid demonstration project: grid regulation ancillary service with a battery electric vehicle, Final report, AC Propulsion, Inc.

[25] Søndergren, C. (2011) Electric vehicles in future market models, Edison Deliverable 2.3, 16 June.

[26] Xu, Z., Østergaard, J. and Togeby, M. (2011) Demand as frequency controlled reserve. *IEEE Transactions on Power Systems*, **26**, 1062–1071.

[27] Xu, Z., Østergaard, J., Togeby, M. and Marcus-Møller, C. (2007) Design and modelling of thermostatically controlled loads as frequency controlled reserve, in *IEEE Power Engineering Society General Meeting*, IEEE, Piscataway, NJ.

[28] Douglass, P.J., Garcia-Valle, R., Nyeng, P. *et al.* (2011) Demand as frequency controlled reserve: implementation and practical demonstration, in *2011 IEEE PES International Conference and Exhibition on Innovative Smart Grid Technologies (ISGT Europe)*, IEEE, Piscataway, NJ.

[29] European Network of Transmission System Operators for Electricity (ENTSOE) (2010) Impact of increased amounts of renewable energy on Nordic power system operation, http://www.stateofgreen.com/Cache/db/db157fbd-4d3b-4350-b3d1-f0e87347d17b.pdf (accessed January 2013).

[30] Akhmatov, V., Gleditsch, M. and Gjengedal, T. (2009) A regulation-caused bottleneck for regulating power utilization of balancing offshore wind power in hourly and quarter hourly based power systems. *Wind Engineering*, **33**, 41–54.

[31] VDE FNN (2011) VDE-AR-N 4105:2011-08 Power generation systems connected to the low-voltage distribution network – technical minimum requirements for the connection to and parallel operation with low-voltage distribution networks. VDE, Frankfurt.

[32] BDEW (2008) Technical guideline on power generation systems on medium voltage networks – guideline for connection and parallel operation of generating plants to medium voltage network, Bundesverband der Energie- und Mittelspannungsnetz, Berlin.

[33] Kirby, B. and Hirst, E. (1997) Ancillary-Service Details: Voltage Control, ORNL/CON-453, Oak Ridge National Laboratory, Oak Ridge TN, December.

[34] Zpryrme (2010) Smart grid insights: V2G, July, http://smartgridresearch.org/standard/smart-grid-insights-v2g/ (accessed February 2013).

9

Operation and Degradation Aspects of EV Batteries

Claus Nygaard Rasmussen,[1] Søren Højgaard Jensen[2] and Guang Ya Yang[1]

[1] Centre for Electric Technology, Department of Electrical Engineering, Technical
University of Denmark, Lyngby, Denmark
[2] Department of Energy Conversion and Storage, Technical University of Denmark,
Roskilde, Denmark

9.1 Introduction

Range has been the Achilles heel of electric vehicles (EVs) compared with cars with internal combustion engines for the last century. The properties of the electrochemical energy storage medium, also known as the battery, determine how much energy can be stored and subsequently transformed into propulsion. Development of the lithium-ion and lithium-polymer battery families has shown significant improvement in gravimetric and volumetric energy capacity during the last 20 years. Here, we show results from laboratory studies of lithium-ion manganese spinel and lithium iron phosphate batteries, and later in this chapter more information about battery module tests can be found.

Comprehensive research programmes on battery technology have been established; for example, the ARPA-E managed by the Department of Energy in the USA. Some of the products envisioned promise an increase in gravimetric energy capacity by a factor of two to four.

A simulation model has been developed at the Technical University of Denmark (DTU) in order to evaluate the effects that use-pattern has on EV battery degradation and lifetime. From a user perspective, the important parameters relating to an EV battery are energy capacity, efficiency and remaining lifetime. These parameters are strongly dependent on a number of variables, such as temperature, depth of discharge (DOD), charge and discharge rates and cycle number. This means that battery use affects battery performance and vice versa. The following aging effects are taken into consideration in the model: DOD, number of cycles, state of charge (SOC), C-rate, temperature, temperature cycling and calendar life.

A battery module test set-up was developed at DTU in order to test a 75 Ah eight-cell lithium–nickel–manganese–cobalt oxide (NMC) Kokam battery module and a 50 Ah 10-cell lithium iron phosphate (LFP) Byd battery module. Results from the battery module tests have been used as inputs for the battery model. The set-up was used to measure the module impedance as well as the impedance of the individual cells in the module. It was observed that when a cell in a module collapses and the cell voltage drops to 0 V due to harsh or prolonged operation of the module, the impedance of the collapsed cell increases dramatically. This means that even though such a cell collapse will have limited impact on the electrical performance of the EV battery pack, local heating of the collapsed cell during pack operation may lead to overheating and subsequent violent collapses of the adjacent cells. This potential safety hazard must be addressed and properly handled by the battery pack's battery management system (BMS).

An equivalent-circuit (EC) model describing the dynamic behaviour of the batteries during operation has been developed. The EC model can be represented as a set of ordinary differential equations and offers a good compromise between accuracy and simulation speed. In relation to the development of the EC model, a technique to find the open-circuit voltage (OCV) as a function of SOC was developed and applied to the batteries. The EC model combines the OCV versus SOC data with SOH data and measurements of the battery impedance to provide real-time information about the battery voltage and SOC as a function of the battery use. The purpose has been to construct a model that describes battery degradation and battery state of health (SOH) as a function of battery use and to construct a model that describes the dynamic behaviour of battery modules as a function of battery use. The objective is to gain knowledge about the overall performance of different battery types for future EVs, as a function of different charging–discharging patterns. Two battery chemistries were tested in order to obtain data for the models: NMC batteries and LFP batteries. The models provide information about different aspects of the battery performance, such as:

- the charge and discharge voltage/current profiles as a function of SOC for the different chemistries;
- the effect of different charging–discharging patterns on cycle lifetime and performance of the batteries.

9.2 Battery Modelling and Validation Techniques

As mentioned above, two battery chemistries has been tested, namely a 75 Ah NMC module from Kokam and a 50 Ah LFP module from Byd. Here, we describe some of the test results and the measurement methods used to characterize the battery modules.

Before degradation testing, the two battery modules were characterized both with charge–discharge cycles at several temperatures (typically 0, 23 and 50 °C) and impedance spectra measurements at various states of charge (typically 25% SOC, 50% SOC and 90% SOC) and various temperatures (typically 0, 23 and 50 °C). Here, we give a detailed description of the characterization method and some of the characterization measurements on the NMC battery module.

9.2.1 Background

During the last decade a rapidly increasing interest in batteries for propulsion in the transport sector has emerged. Since battery performance is drastically affected by the operation pattern, it is important to characterize the battery degradation as a function of the operation pattern. This is not a trivial task, since the battery performance is not a measurable quantity in itself, but covers several aspects such as the internal resistance and the capacity. Further, the operation pattern covers several aspects such as the DOD, SOC, C-rate, temperature and calendar life.

Thus, detailed characterization of battery modules is necessary to construct reliable models that incorporate performance-related aspects of the modules, such as thermodynamics, electrochemical reaction kinetics and degradation mechanisms. Charge–discharge curves, temperature measurements and battery impedance measurements can provide detailed information about these aspects. Charge–discharge curves can be used to measure the battery OCV and the internal resistance. Temperature measurements provide information about the thermodynamic reactions, and impedance spectra yield detailed information about the reaction kinetics.

The battery module wear due to specific operation patterns can be analysed with micro-cycle tests where a single operating parameter, such as the battery temperature, is varied slightly while keeping the other operating parameters as constant as possible; for example, the charge–discharge cycles are kept constant. The internal resistance, the capacity and the impedance of the cells and the module can be measured before and after each series of micro-cycles. In this way it is possible to map the battery module wear as a function of the various operating conditions and thus to develop a model that predicts the module wear due to a more complex operating pattern, as described by Safari *et al.* [1].

Battery impedance spectra provide valuable knowledge about the reaction kinetics of physical/chemical processes taking place inside the battery. For this reason, the impedance spectra can provide information that can be used to develop improved SOH models and to describe how the battery voltage (or current) responds to changes in the current (or voltage) [2, 3].

It is usually very time consuming to measure impedance spectra on batteries because some of the investigated electrochemical processes occur very slowly. This is in particular true for single-sine methods, where the impedance is measured at a single frequency at a time. With time-domain methods (TDMs) it is possible to simultaneously measure the impedance at all frequencies within the measurement range. The minimum frequency that accurately can be analysed by a TDM measurement is the inverse of the measurement time [4]. At high frequencies, the measurement accuracy of TDM is limited by the time resolution of the voltage or current perturbation and the time resolution and accuracy of the measurement of the resulting current or voltage response from the battery. Normally, this limits the maximum frequency that can be analysed by TDM below 100 Hz. For this reason it is interesting to combine high-frequency single-sine measurements with low-frequency TDM measurements to conduct fast and accurate impedance measurements in a broad frequency range.

It should be noted that multi-sine techniques exist [5], which can decrease the measurement time by approximately a factor of four compared with single-sine measurements [6]. Further, real-parallel impedance solutions like the 'PAD4' by Zahner can measure impedance on several cells simultaneously. Unfortunately, the equipment for both multi-sine and real-parallel techniques is generally more expensive than the equipment used for the measurement technique presented.

Several techniques to convert TDM measurements into frequency-domain measurements exist [4, 7, 8]. Klotz *et al.* have recently presented a study on combined TDM and electrochemical impedance spectroscopy (EIS) measurements of a lithium-ion cell with an internal resistance on the order of 0.5 Ω [9]. Here, we present results on combined TDM and EIS measurements, but we use the TDM method to simultaneously obtain impedance spectra on eight cells in a battery module, and the cell impedances are on the order of 1 mΩ.

The transformation of the TDM measurements into the frequency domain presented uses an EC based on four (RC) circuits in series with a resistor and a capacitor, where (RC) is a parallel connection between a resistor and a capacitor. The EC is used to model the overvoltage as a function of time due to a step current of 1.0 A, and the values of the resistors and capacitors obtained from the modelling are used to calculate the impedance. The number of (RC) circuits was chosen to be the smallest number that still allowed an accurate fit of the overvoltage.

9.2.2 Experimental Testing Techniques

The battery module tested is an SLPB125255255H_8S1P module supplied by Kokam Co. Ltd. The module consists of eight serially connected 75 Ah cells with NMC cathodes and lithium–carbon anodes. The cell electrolyte consist of lithium salt (e.g. LiPF$_6$), organic solvent (such as ethylene carbonate), gel polymer and performance/safety enhancing additives. The cells are usually referred to as NMC cells and Kokam supplies the following technical specification for the NMC cells: the minimum discharge cell voltage is 3.0 V, the nominal voltage is 3.7 V and the maximum charge voltage is 4.15 ± 0.03 V. The module is 29.3 cm high, 27.5 cm wide and 10.5 cm long.

The battery module voltage, the individual cell voltages, the current and the module temperature were consecutively logged with a Keithley 2750 system with two Keithley 7700 cards and a Keithley 7702 card. A thermistor placed between cell 1 and cell 2 was used to measure the temperature inside the module. The thermistor was placed 1.5 cm below the top and 10 cm from one of the sides of the module.

The module was charged with two Delta-Elektronika SM 60-100 power supplies in parallel and discharged with an E-load EA9080-600. The power supplies and E-load could be decoupled from the main electric circuit with power relays (not shown in Figure 9.1). The module and cell impedance was measured with normal single-sine impedance spectroscopy using a Solartron 1252A. A Kepco BOP 50-4M was used to boost the 16 mA AC current from the Solartron to 1 A AC. When current passes the current transducer LEM LA-125 it results in current passing through the measurement resistor R_m and accordingly a voltage drop across R_m. On dividing this voltage drop by the current passing through the current transducer, the 'resistance' of the combined current transducer and R_m is obtained. Using an AC current instead of a DC current, we can measure the impedance of the combined current sensor and measurement resistor. We call this impedance Z_{LEM}. During the impedance measurements the AC voltage drop across R_m was measured with 'V2' on the Solartron 1252A.

The current passing through the battery module changes the voltage of the individual cells and of the module as a whole. The cell voltages or the module voltage was measured with 'V1' on the frequency analyser. The coax-cables from the Solartron to the individual cell electrodes were multiplexed with a Keithley 2750 and two Keithley 7700 cards to enable automated measurements without needing to move the impedance cables.

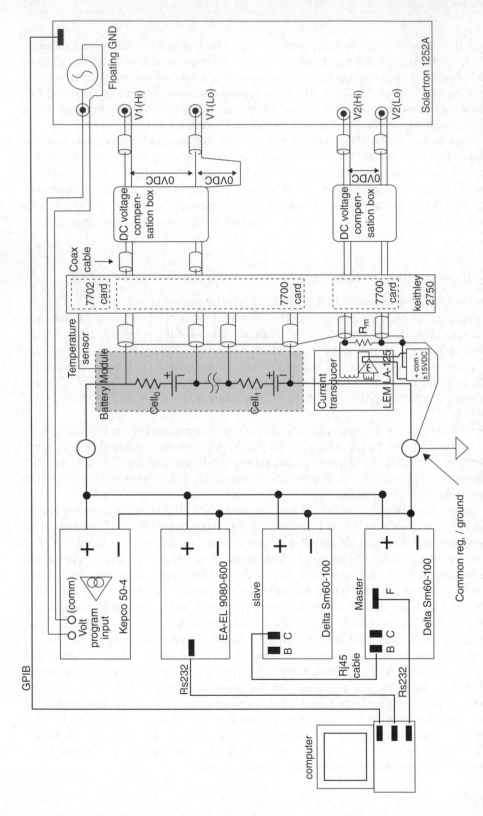

Figure 9.1 Diagram of the wiring between the most important parts of the test system. Power supplies: Delta SM60-100; E-load: EA-EL 9080-600; current sensor: LEM LA-125 with measurement resistor R_m; frequency analyser: Solartron 1252A; multiplexer: Keithley 2750; battery module: Kokam SLPB12525255H_8S1P.

Prior to the impedance measurements, the DC voltage compensation boxes were automatically adjusted so that the DC voltage from the module or one of the cells was removed. This means that only the AC voltage is transferred to the frequency analyser in order to use the 5 V common-mode rejection of the Solartron 1252A and to make use of the most sensitive measurement range of the frequency analyser.

V1/V2 was recorded with the Solartron 1252A in the frequency range 2.6 kHz to 0.6 mHz. The impedance of the battery was obtained from

$$Z_{Bat} = \frac{V1}{V2} Z_{LEM} \tag{9.1}$$

At high AC frequencies the power supplies (Delta SM60-100) and the E-load (EA-EL 9080-600) stop acting as perfect galvanostats and significant error currents pass through these devices. However, the current transducer is in series with the battery module, so it only senses the current passing the battery module. The AC voltage generated in the battery module and its cells is caused by the AC current applied to the battery. Thus, the impedance measurement is not affected by the leak currents in the E-load and power supplies.

The module and cell impedances were also measured with a Laplace transform of the module and cell overvoltage. The overvoltages were measured as a function of time after onset of a step current of 1 A. The data acquisition lasted 1 h and the data was measured every 2 s. The Kepco BOP 50-4M was used to apply the step current for the Laplace measurements.

The module charge capacity was tested with a charge–discharge curve ranging from the manufacturer's maximum charge voltage to the minimum discharge voltage. We define the module to be fully charged when the voltage of one of the cells in the battery module reaches 4.15 V at $0.13C$, which corresponds to a charge current of -10 A. The SOC of the fully charged battery module is defined as 100% SOC. The fully discharged battery module (i.e. when one of the cells in the battery module reaches 3.0 V at $0.13C$, i.e. at a discharge charge current of 10 A) is defined as 0% SOC.

SOCs between 0% and 100% are measured by Coulomb counting; that is, by subtracting the net charge flow from the fully charged module divided by the measured module charge capacity.

The module OCV is measured as a function of SOC by a method previously described by Abu-Sharkh and Doerffel, who used it for fast OCV measurements at high C-rates [10]. Here, we measure the OCV versus SOC from a charge–discharge cycle which consists of a series of discharge steps from 100% SOC to 0% SOC followed by a series of charge steps from 0% SOC to 100% SOC. Each step lasts 6 min and consists of 5 min of either charging or discharging at $0.13C$ (i.e. at either -10 A or 10 A) and 1 min at $0C$ (i.e. at 0 A). The voltage measurement during the $0C$ periods in the charge steps and discharge steps is used to determine the OCV versus SOC. This is described in further detail below.

Nonideal coulombic efficiency can lead to large accumulated errors for this type of SOC measurement if the battery is cycled several times without reaching a fully charged or discharged condition where recalibration of the SOC is possible. Here, we present a charge–discharge curve from 100% SOC to 0% SOC and back to 100% SOC. This SOC definition does not fully account for self-discharge with time. However, the self-discharge of the module prior to any degradation tests was approximately 1 mV per day and can thus be ignored during the charge–discharge curves presented here, which lasted for approximately 20 h.

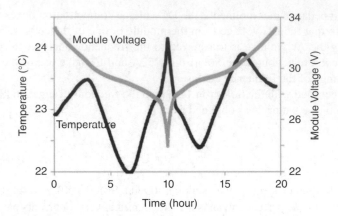

Figure 9.2 Voltage and temperature measured as a function of time during a charge–discharge cycle of a 75 Ah NMC battery module.

Battery Module Measurements

The battery module temperature and voltage were measured as a function of time during a full charge–discharge cycle; see Figure 9.2. Before the charge–discharge cycle, the module was charged to 100% SOC. This was done in steps of 6 min. In each step, the battery was first charged with -10 A for 5 min and the current was reduced to 0 A for 1 min. As soon as one of the eight cells in the module reached the maximum charge voltage of 4.15 V, the current was interrupted and the current SOC was defined as 100% SOC.

Immediately after this, the module was discharged with 10 A for 5 min. Then the battery current was set to 0 A for 1 min. This was repeated until one of the eight cells in the module reached 3 V. After this, the module was charged in steps of 5 min with 10 A and 1 min at 0 A until one of the eight cells reached 4.15 V (100 % SOC).

The battery capacity was measured as the current multiplied by the total time during discharging. The battery capacity was found to be 75.3 Ah when the ambient temperature was 22 °C. The temperature of the battery module at the beginning of the charge–discharge cycle was higher than the ambient temperature due to pre-charging to reach 100% SOC.

Immediately after this, the module was discharged with 10 A for 5 min. Then the battery current was set to 0 A for 1 min. This was repeated until the one of the eight cells in the module reached 3 V. After this, the module was charged in steps of 5 min with 10 A and 1 min at 0 A until one of the eight cells reached 4.15 V (100% SOC).

The battery capacity was measured as the current integrated over time during discharging. The initial capacity of the battery module was found to be 75.3 Ah when the ambient temperature was 22 °C. The initial temperature of the battery module was higher than the ambient temperature due to pre-charging to reach 100% SOC. Note how the battery heats up during the first 3 h of discharging and cools down subsequently. This is thought to be due to entropy changes arising from structural transformations in the Li–C anode and NMC cathode phase changes [7, 8, 9, 10].

The module voltage as a function of SOC is shown in Figure 9.3. The grey line shows the module voltage during the 5 min periods at 0.13C. The black points in the figure show the

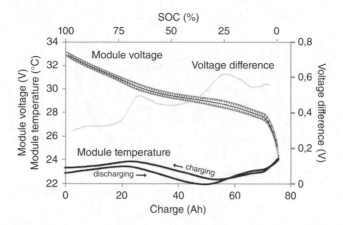

Figure 9.3 Charge–discharge curve for the 75 Ah NMC battery module. The charge–discharge curve is made as a series of discharge steps followed by a series of charge steps. Each step takes 6 min: 5 min at $0.13C$ and 1 min at $0C$. The grey line shows the module voltage during $0.13C$. The black points show the module voltage during $0C$. The thin black line shows the SOC calculated from the black points. The grey points show the voltage difference between the upper and lower grey lines. The thick black line shows the module temperature.

module voltage just before the end of the 1 min periods at $0C$. As described by Abu-Sharkh and Doerffel [10], an estimate of the OCV at a given SOC can be obtained as the average voltage of the upper and lower black points. If the periods at $0C$ had been sufficiently long, the module voltage would relax until it reaches OCV. The thin black line in Figure 9.3 shows this calculated OCV estimation. The difference between charging and discharging voltage as a function of SOC is shown with grey dots. The voltage was measured immediately before the 1 min current interruptions in the charge and discharge steps. In the figure, the module temperature is also shown as a function of SOC with a thick black line.

The module impedance was measured at 90% SOC. This was done both in an ordinary single-sine measurement with a Solartron 1252A and with a Laplace-transformed overvoltage curve. The overvoltage as a function of time after onset of a step current of 1.0 A is shown in Figure 9.4. The overvoltage is measured with a frequency of approximately 0.5 Hz for 1 h.

The criteria for the model of the overvoltage in Figure 9.4 are threefold. First, the model should fit the data in the time domain as well as possible. From the inset in Figure 9.4, which shows the residual between the measurement data and the model data, it is seen that this is indeed fulfilled by the chosen model. Second, it should be possible to analytically transform the model into the frequency domain. Third, the model should provide impedance data that are in good agreement with the single-sine measurements. We do not assign a physical meaning to the individual variables in the model of the overvoltage. However, the sum of resistances in the model is described as the internal resistance of the battery module as presented below.

The model of the overvoltage is obtained from an equivalent circuit, $R_s C_{bat}$ $(RC)_1 (RC)_2 (RC)_3 (RC)_4$, where R is a resistor and C is a capacitor, (RC) is a parallel connection between a resistor and a capacitor and R_s, C_{bat} and $(RC)_{1-4}$ are serially connected. As shown in the 'Discussion of results' section, the number of (RC) circuits was chosen as the smallest number that fulfilled the first of the above-mentioned criteria.

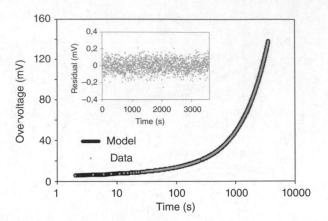

Figure 9.4   Overvoltage measured on a 75 Ah NMC battery module as a function of time after onset of a -1.0 A current. The grey dots are the measured data. The black line is a model of the overvoltage. The inset shows the residual difference between the model and the measured data. We use the model to derive the low-frequency impedance of the battery module.

The current in the time domain can be written as $I(t) = I_0 u(t)$, where I_0 is the current amplitude (i.e. -1.0 A) and $u(t)$ is a step function which is zero when $t < 0$ and one when $t \geq 0$. The Laplace transform of $I(t)$ is $I(s) = I_0/s$, where $s = j\omega$, j is the complex unity and ω is the angular frequency. The overvoltage in the frequency domain (or s-domain), $U_m(s)$, is the product of $I(s)$ and the impedance of the EC $Z_m(s)$; that is:.

$$U_m(s) = I(s)Z_m(s)$$
$$= \frac{I_0}{s}\left(R_s + \frac{1}{sC_{bat}} + \frac{R_1}{1 + sR_1C_1} + \frac{R_2}{1 + sR_sC_2} + \frac{R_3}{1 + sR_3C_3} + \frac{R_4}{1 + sR_4C_4}\right) \quad (9.2)$$

The overvoltage in the time domain $U_m(t)$ is given as the inverse Laplace transform of $U_m(s)$:

$$U_m(t) = u(t)I_0\left\{R_s + tC_{bat} + R_1\left[1 - e^{-t/(R_1C_1)}\right] + R_2\left[1 - e^{-t/(R_2C_2)}\right]\right.$$
$$\left. + R_3\left[1 - e^{-t/(R_3C_3)}\right] + R_4\left[1 - e^{-t/(R_4C_4)}\right]\right\} \quad (9.3)$$

Values were assigned to the variables in the expression of $U_m(t)$ by fitting the model of $U_m(t)$ to the data in Figure 9.4; that is, by minimizing the error sum:

$$\text{Err} = \sum_{t_n} \frac{1}{t_n}(U_m(t_n)) - U_d(t_n))^2 \quad (9.4)$$

where t_n is the time after onset of the step current when the nth measurement point was recorded. $U_d(t_n)$ are the data points shown in Figure 9.4.

A weighing factor of $1/t_n$ was used in the calculation of Err to equalize the weighting of the variables involved as much as possible and thus stabilizing the conversion of the fitting routine: Suppose $(RC)_1$ primarily models the part of the impedance spectrum from 0.1 to 0.01 Hz and $(RC)_4$ primarily models the part of the impedance spectrum from 0.001 to 0.0001 Hz. This means the $U_m(t)$ measurements from 10 to 100 s are primarily modelled by $(RC)_1$ and the $U_m(t)$ measurements from 1000 to 10 000 s are primarily modelled by $(RC)_4$ (temporarily disregarding that we only measured $U_m(t)$ for 3600 s). Because $U_m(t)$ was measured approximately every other second, 45 measurements were conducted from 10 to 100 s and 4500 measurements were obtained between 1000 and 10 000 s. This means that

$$(\sum_{t_n=10\,s}^{100\,s} \frac{1}{t_n} \approx \sum_{t_n=1000\,s}^{10\,000\,s} \frac{1}{t_n} \approx \frac{1}{2}\ln(10) \tag{9.5}$$

which means the variables in $(RC)_1$ and $(RC)_4$ are approximately equally weighted when we minimize the expression in (9.4).

In order to stabilize the modelling routine we fix the value of R_s using the following relation derived from equation (9.3):

$$R_s = \frac{U_m(t_1)}{I_0} - t_1 C_{bat} - R_1\left[1 - e^{-t/(R_1 C_1)}\right] - R_2\left[1 - e^{-t/(R_2 C_2)}\right]$$
$$-R_3\left[1 - e^{-t/(R_3 C_3)}\right] - R_4\left[1 - e^{-t/(R_4 C_4)}\right] \tag{9.6}$$

where $U_m(t_1)$ is the overvoltage measured at t_1 and t_1 is the time after onset of the step current where the first measurement of the overvoltage occurs.

The values of the remaining variables C_{bat}, R_1, C_1, R_2, C_2, R_3, C_3, R_4 and C_4 obtained from the fitting routine were used to calculate $Z_m(s)$, and the result is shown in Figure 9.5 (legend: 4 (RC)) together with the battery module impedance measured with the single-sine measurement method. From the figure it is seen that the two methods yield reasonably accurate data below 60 mHz. Above 60 mHz the Laplace method measurements (not shown) increasingly deviate from the single-sine measurements possibly due to the limited data acquisition frequency. The figure also shows $Z_m(s)$ obtained from modelling the overvoltage using an EC having one, two and three (RC)s in series with a capacitor and a resistor.

In order to measure the changes in the electrode kinetics of the individual cells of the module, we measured impedance spectra on the individual cells in the battery module. Just as we did with the measurements of the battery module impedance spectra shown in Figure 9.5, the spectra were measured both with the single-sine measurement method and with the TDM measurement method. Again, R_s was fixed in order to assure a stable conversion of the fitting routine using the expression in (9.6). The impedance data obtained are shown in Figure 9.6.

Discussion of Results

The first part of the discussion section examines the thermal behaviour of the battery module presented in Figure 9.2 and Figure 9.3.

The changes in the exothermal/endothermal behaviour of the battery module are thought to be due to entropy changes arising from structural transformations in the Li–C anode and NMC cathode phase changes [11, 12, 13, 14] which in turn lead to the observed changes of

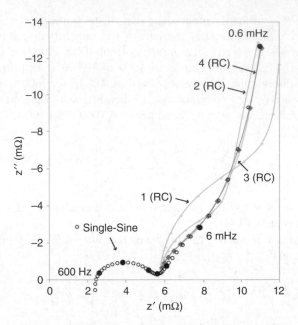

Figure 9.5 Impedance measured on the 75 Ah NMC module. The impedance is measured both with a single-sine method and a Laplace method (9.3) with one, two, three and four (*RC*)s. The single-sine measurements shown with solid black markers denote the frequency decades.

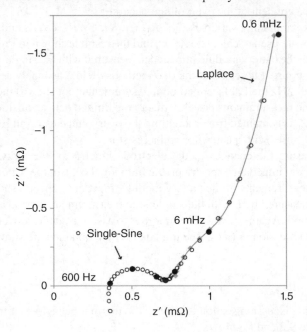

Figure 9.6 Impedance measured on a single cell in the 75 Ah NMC module. The impedance is measured both with a single-sine method and a Laplace method. The single-sine measurements shown with solid black markers denote the frequency decades. The two methods overlap reasonably well at frequencies below 60 mHz.

$$SOC_1 \; > \; SOC_2 \; > \; SOC_3$$

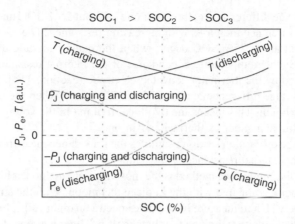

Figure 9.7 A description of the observed local minima in the battery temperature during charging and discharging shown in Figure 9.3 by Joule heat and changes in the entropy of the electrochemical processes. The local maxima in the battery temperature can be described in a similar manner.

the temperature inside the battery module. In Figure 9.7, this is exemplified by schematic data for two local minima of the battery temperature as observed in Figure 9.3. At SOC_2, the entropy of the electrochemical processes in the battery changes sign. The heat power P_e from the electrochemical processes can be described as

$$P_e = \frac{I}{nF} T \Delta S \tag{9.7}$$

where I (A) is the current, n is the number of electrons involved in the electrochemical process, F is the Faraday constant, T is the temperature and ΔS is the entropy change. Note that P_e changes sign with the sign of the current. The Joule heat due to current passing through the electrolyte is given as

$$P_J = R_i I^2 \tag{9.8}$$

where R_i is the electrolyte resistance. P_J is positive during both charging and discharging. The change in the battery temperature per unit time can be described as

$$\frac{dT}{dt} = \frac{P_e + P_J}{C_p} \tag{9.9}$$

where C_p is the heat capacity of the battery. At SOC_1, during discharge $-P_e = P_J$. At $SOC > SOC_1$, during discharge $P_e + P_J < 0$ and the battery cools down. At $SOC < SOC_1$, during discharge $P_e + P_J > 0$ and the battery heats up. At SOC_3, during charge $-P_e = P_J$. At $SOC < SOC_3$, during charge $P_e + P_J < 0$ and the battery cools down. At $SOC > SOC_3$, during discharge $P_e + P_J > 0$ and the battery heats up.

The module voltage presented in Figure 9.2 was used to calculate the voltage difference between charging and discharging as shown by the grey points in Figure 9.3. Here, two local maxima for the voltage difference are observed: one at about 25% SOC and one at about 75% SOC. The voltage difference increases with increasing resistance. The local maximum of the voltage difference at about 25% SOC occurs when the battery module temperature reaches a local minimum. Such a temperature dependence of the internal resistance is also expected and can be explained by the thermally activated electrode reactions [14]. However, the local maximum of the voltage difference at about 75% SOC occurs when the battery temperature reaches a local maximum. This means the two local minima in the voltage difference cannot be explained in a simple manner by thermal activation.

The next part of the discussion section examines the impedance measurements and provides an expression for the internal resistance of the battery.

As stated above, the single-sine method is the most accurate method at high frequency but is slow at low frequencies compared with the Laplace technique. With the single-sine technique it takes approximately 30 h to measure impedance spectra on eight cells plus the module down to 0.6 mHz. With the Laplace technique it only takes 1 h. The remaining data from 60 mHz to 2610 Hz measured with the single-sine method on eight cells plus the module takes less than 30 min. For this reason, a much faster data acquisition can be obtained together with a good precision in the entire frequency range when the two measurement techniques are combined. Thus, in the current example it is possible to reduce the data acquisition time by a factor of 20 compared with ordinary single-sine measurements.

Equation 9.3 describes how the battery voltage evolves with time due to a step current. If the amplitude of the step current is sufficiently small and the battery is charged or discharged for a sufficiently long time, we can neglect the transients and ignore changes in the impedance due to changes in SOC. Then, (9.3) reduces to

$$U_m(t) = u(t)I_0(R_s + tC_{bat} + R_1 + R_2 + R_3 + R_4) \tag{9.10}$$

This means the voltage difference ΔU between the battery voltage measured at the same SOC during charging and discharging can be obtained from (9.10) as

$$\Delta U = U_m^c(t) - U_m^d(t) = 2I_0(R_s + R_1 + R_2 + R_3 + R_4) \tag{9.11}$$

where $U_m^c(t)$ and $U_m^d(t)$ are the battery voltages measured at the same SOC during charging and discharging respectively. Thus, the internal resistance as a function of SOC can be measured from a very slow constant-current (CC) charge–discharge curve as $\Delta U/2I_0$.

In relation to (9.3), the number of (RC)-circuits was chosen to be the smallest possible number that provides a good fit between the measured overvoltage and the modelled overvoltage. A generic expression for the internal resistance R_i of the battery would be

$$R_i = R_s + \sum_{n=1}^{N} R_n \tag{9.12}$$

where N is the number of (RC)-circuits.

We have modelled the overvoltage in Figure 9.4 with an expression like the one presented in (9.3) but with one, two, three and four (RC)s; that is, $N = 1, 2, 3, 4$. Naturally, the difference between the measured and calculated overvoltages decreases with N, as seen from Figure 9.4.

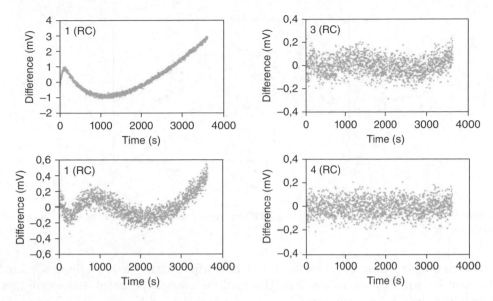

Figure 9.8 The residual between the measured and calculated overvoltages is shown in Figure 9.4 using an expression similar to (9.3) but with a varying number of (RC)-circuits. The 4 (RC) figure is similar to the inset in Figure 9.4. The corresponding impedance spectra are shown Figure 9.5.

In agreement with this, the single-sine measurements and Laplace-method measurements increasingly overlap with increasing number of (RC)-circuits, as seen from Figure 9.8.

Ideally, the internal resistance given by (9.12) should convert asymptotically to a given value for an increasing number of (RC)-circuits in the model. The values of R_i from the modelling are shown in Table 9.1. Also, the smallest characteristic frequency of the model (RC)-circuits are shown in the table. The overvoltage measurement was conducted for 1 h corresponding to a characteristic frequency of 0.28 mHz. The tabulated values for the characteristic frequency are comparable to, or even smaller than, this frequency.

An exact determination of R_i requires that the overvoltage measurement time is significantly longer than the characteristic time of the transients modelled by the (RC)s. If this had been the case, the inverse of ω_0 would have been smaller than the measurement time. At the same time, the step current that causes the overvoltage needs to be sufficiently small to ensure a stable measurement of the battery impedance; that is, that changes in the battery impedance due to the change in SOC during the overvoltage measurement can be ignored.

Table 9.1 Internal resistance and lowest characteristic frequency as a function of the number of (RC)-circuits.

No. of (RC)-circuits	R_i (mΩ)	ω_0 (mHz)
1	12.3	2.5
2	15.9	0.31
3	17.7	0.18
4	19.4	0.13

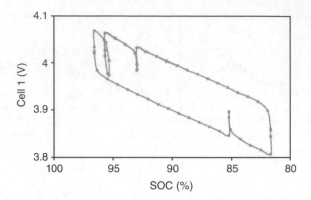

Figure 9.9   Micro-cycle of a single cell in a NMC battery module; 238 micro-cycles was executed at 40 °C and 50 A.

In the example presented, at least the first of these requirements was not fully satisfied and this explains why R_i does not converge. However, the example presented demonstrates the principles of how R_i can be measured.

9.2.3   Degradation of Battery Modules

The aim of the experimental work was to characterize NMC and LFP battery modules. This can be accomplished by a series of micro-cycle tests. A micro-cycle test is a test where one of the operating parameters of the battery module is slightly varied while all other operating parameters are kept as constant as possible. By measurements of the change in charge capacity, internal resistance and impedance of the battery modules due to the micro-cycle tests, it is possible to identify operating conditions that severely wear the battery modules. Further, it is possible to use the micro-cycles to create a model that predicts the battery wear as a function of battery usage history as described by Safari *et al.* [1]. An example of a micro-cycle test is shown in Figure 9.9. A total of 238 of these micro-cycles at 40 °C, 50 A were performed on the NMC battery module.

Charge–discharge curves were measured before and after the 238 cycles. As seen from Figure 9.10, the charge capacity of the module clearly decreased due to the micro-cycling. The internal resistance is also seen to increase due to the 238 micro-cycles, since the voltage difference between the charge voltage and discharge voltage at a given SOC is larger after the cycles than before the cycles. The internal resistance can be measured as the voltage difference between the charge voltage and discharge voltage at a given SOC, divided by the difference between the charge current and the discharge current – as explained in Section 9.2.2.

Similar micro-cycle tests were applied to the LFP module (Figure 9.11). It was tested with 172 micro-cycles at 20 °C, 25 A. The module was subsequently characterized with a charge–discharge cycle. As seen from the impedance spectra in Figure 9.9, the impedance increased during the micro-cycle tests. The spectra were measured on two of the cells in the module before and after the micro-cycle tests and they were measured at 23 °C at 90% SOC. The data obtained from the micro-cycle tests on the two battery modules can be used as experimental input to construct the battery models.

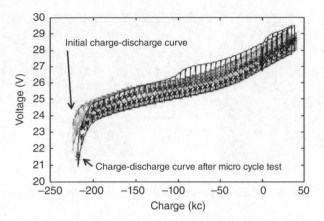

Figure 9.10   Charge–discharge curves on an NMC battery module before and after 238 micro-cycles as shown in Figure 9.9. The grey curve shows the charge–discharge curve before the micro-cycle tests and the black curve shows the charge–discharge curve after the micro–cycle tests. Crosses show the cell voltage during 1 min resting periods during the charge–discharge cycles. Note how the battery capacity has decreased and how the internal resistance, in particular at higher SOC (high voltage), has increased.

9.2.4   Test Set-Up and Results

The laboratory test-bench for battery testing consists of an LFP battery pack, where each battery cell has a nominal voltage of 3.2 V and capacity of 40 Ah. A BMS is deployed for monitoring cell status in the battery pack, including voltage, current, temperature and SOC. The charging process is handled by a three-phase charger. The BMS transmits alarm signals to the charger in the case of battery overvoltage, undervoltage, overheating and so on. The charger will then automatically reduce the charging current or voltage and eventually turn off the process.

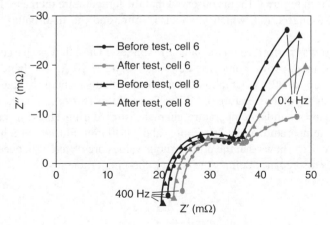

Figure 9.11   Impedance spectra measured on two cells of a BYD module before and after 172 micro-cycles. The frequency range is from 1 to 400 Hz.

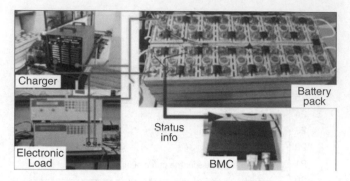

Figure 9.12   Laboratory configuration for battery characterization.

The discharge process of the battery is handled by an electronic load with manual or programmable capability. The electronic load is most often used in CC mode. In this mode the current is only adjusted at the beginning and then remains constant until the end of testing. A minimum voltage condition can also be applied to protect the battery. The whole test set-up is shown in Figure 9.12.

In addition, a temperature probe and several thermistors are available to monitor the temperature of different cells during testing of the battery pack. PC software is used to log real-time data of battery status, with minute resolution.

Temperature Measurements

The temperature behaviour of an LFP cell has been measured and recorded with a CC discharging cycle using the electronic load of Figure 9.12, with current fixed to 40 A. The LFP cell was discharged from SOC of 80% down to 2.5 V, which is the minimum recommended discharge voltage claimed by the manufacturer. The test was conducted at an ambient temperature of 22 °C. From Figure 9.13 it is observed that the temperature increases linearly through the whole discharging period, which is about 52 min, and the final battery temperature is about 37 °C.

A charging test was performed on the same LFP battery. The cell was charged using a power supply set at CC mode, with a charging current of $0.5C = 20$ A. The battery was charged from 20% SOC, up to 4.25 V, which is the maximum battery charge voltage. The test was conducted at an ambient temperature of 23 °C. The result is shown in Figure 9.14.

Figure 9.14 shows similar temperature characteristics during charging and discharging, obtaining a final temperature about 36 °C after about 140 min. Further tests include a battery OCV test. Figure 9.15 shows that the ampere-hour values are different between charging and discharging. This discrepancy defines the battery charging efficiency:

$$\eta = \frac{\int I_{\text{discharge}}\, \mathrm{d}t}{\int I_{\text{charge}}\, \mathrm{d}t} \qquad (9.13)$$

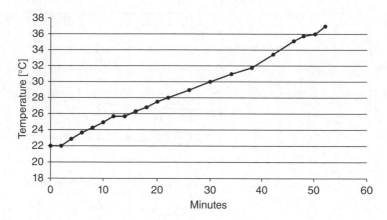

Figure 9.13   Temperature of an LFP cell during discharge at a rate of 1*C*, at an ambient temperature of 22 °C.

9.3   Thermal Effects and Degradation of EV Batteries

The work presented here is based on a model study aimed at obtaining a better understanding of EV battery performance and degradation effects, amongst which thermal effects are central. The model is intended to serve as a simulation tool, as well as a data collection on degradation effects.

It has been the aim of this model to include degradation effects which may be quantified by the use of degradation measurements and to enable a study of the influence that vehicle-to-grid (V2G) operation will have on lifetime and performance of EV batteries. An initial analysis of such operation has also been made.

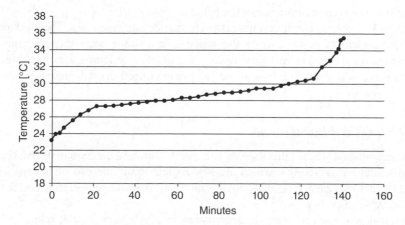

Figure 9.14   Temperature of an LFP cell during charge at a rate of 0.5*C*, at an ambient temperature of 23 °C.

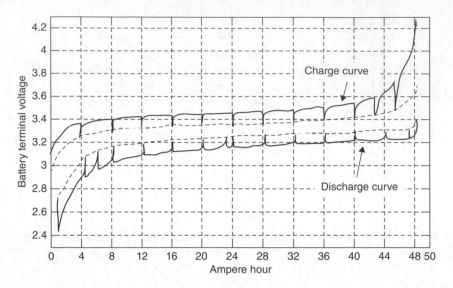

Figure 9.15   Cell voltage as function of injected/extracted charge.

9.3.1   Introduction to Battery Degradation

The prospects of using EV batteries as buffer or storage units in the electric power system have been described and investigated [15]. This raises the question of how this will affect battery life, and for that purpose a simulation model can be very useful in order to evaluate the effects that use-pattern has on EV battery degradation, lifetime and performance.

From a user perspective, the important parameters relating to an EV battery are, for example, remaining energy capacity, remaining lifetime and efficiency. These parameters are strongly dependent on a number of variables, such as temperature, DOD, charge and discharge rates and cycle number. All of these effects, as well as several others, add up in a complex manner to result in the observed performance of the battery.

This means that battery use affects battery performance and vice versa. The work presented here has been aimed at creating a model that will enable a closer study of degradation effects, amongst which the thermal effects are central. The model incorporates a user interface and is intended to serve as a simulation tool, as well as a data collection on degradation effects.

9.3.2   Theoretical Background

It has been established [16, 17] that the energy capacity fade and increase in internal resistance of the individual battery cells – which are key indicators of the use of battery life – are dependent on factors such as charge/discharge rate (C-rate), DOD, temperature and the number of charge–discharge cycles.

The physical processes resulting in degradation are, to a large extent, related to the cell electrodes. The charging and discharging processes lead to phase changes within the electrodes, which in turn result in inhomogeneous volume changes. These volume changes cause microscopic voids and cracks which in turn lead to a reduction in the overall electrode contact area. This is observed as capacity fade and increase in internal resistance of the cells. Temperature variations will tend to have similar effects due to differences in thermal expansion coefficient between different phases. The following aging effects have so far been taken into consideration in the model presented here:

- DOD
- number of cycles
- SOC
- C-rate
- temperature and temperature cycling
- time.

The cycle number and DOD of the individual cycles are interconnected as a battery degradation mechanism. In order to model this, a so-called rain-flow-counting algorithm [18] has been implemented in the model. This algorithm counts and sorts charge cycles according to DOD, after which a lifetime consumption weight factor is attributed to each cycle.

The rain-flow-counting approach to estimating degradation is traditionally used when calculating mechanical degradation. In the case of battery cells, the degradation mechanisms are similar [19] and this method, therefore, is expected to provide a realistic picture of the degradation process.

Thermal Characteristics

A thermal model of an EV battery requires thermal properties of the individual cells to be determined. A lithium-ion battery cell was therefore tested in order to determine its thermal properties. This was done by placing a resistive heat source on one side of the cell and measuring the transient temperature response on the other side of the cell. A simulation model was then calibrated against the measurement results to obtain the relevant values.

A fairly accurate description of the thermal characteristics of a battery cell (or collection of battery cells) is provided by

$$mC\frac{dT}{dt} + \frac{1}{R_{th}}(T - T_a) = P \qquad (9.14)$$

with m being the cell mass and C the thermal capacity of the cell material. T_a is the ambient temperature and P is the power dissipated within the cell due to a current I running through the internal resistance R_i:

$$P = I^2 R_i \qquad (9.15)$$

R_{th} (K/W) is the thermal resistance between the cell (or battery) interior and ambient. Calculating the value of R_{th} is not a straightforward task, but it can be attributed to a thermal resistance of the battery material as well as a thermal resistance between the battery surface and ambient. For heat moving through a wall, R_{th} is given as

$$R_{th} = \frac{d}{\lambda A} + \frac{1}{h A} \tag{9.16}$$

with d being the wall thickness, λ is the thermal conductivity of the wall material, A is the surface area and h is the heat transfer coefficient of the surface.

Solving (9.14) leads to an exponential equation. A change in ambient temperature or dissipated power of the battery will result in a transient response with a temperature distribution that exponentially approaches a steady-state value:

$$T(t) = \exp\left(\frac{-t}{\tau}\right)(T_0 - T_\infty) + T_\infty \tag{9.17}$$

with T_0 being the initial temperature and T_∞ the steady-state temperature. The parameter τ is a thermal time-constant determined by the material parameters and geometry of the cell. The tested cells were found to have a thermal conductivity of $\lambda \approx 0.4$ W/(m K) and a thermal capacity of $C \approx 1400$ J/(kg K). For a single cell of the type tested here, this leads to a time constant in the range of

$$\tau \sim \frac{\rho C}{\lambda}\left(\frac{Vd}{2A}\right) \approx 1 \text{ h} \tag{9.18}$$

with ρ being the cell material density. The cell geometry is determined by volume V, surface area A and thickness d of the cell. For a large collection of battery cells the time constant becomes even longer, and this means that thermal equilibrium will usually not be obtained during 'normal' EV operation. Transient modelling, therefore, is essential when describing the battery thermally. Figure 9.16 shows how a thermal model has been adjusted to fit the results of a test and thereby determine the thermal properties of a battery cell.

The thermal part of the model is based on the finite-difference method, with the battery cells being placed in a two-dimensional array. Each battery cell constitutes a thermal element, as illustrated in Figure 9.17. This means that possible hot spots within the individual cells are not included in the analysis.

Each cell is seen as a homogeneous medium with a certain thermal capacity and thermal conductivity. Each cell is also assumed to contain a homogeneous heat source dissipating power due to the internal resistance of the cell R_i, which is typically in the range of a few milli-ohms.

In addition to Joule heating resulting from the internal resistance, a reversible heating effect may also be taken into account [20]. The reversible heat per mass unit is given as

$$q_{rev} = T \Delta S = n F T \frac{dV_{oc}}{dT} \tag{9.19}$$

with n being the number of electrons per molecule of reactant, F (C/kg) is the Faraday constant, T is temperature and V_{oc} is the OCV. Depending on the type of battery, dV_{oc}/dT may be either positive or negative; if $dV_{oc}/dT > 0$, the effect will heat when charging and cool

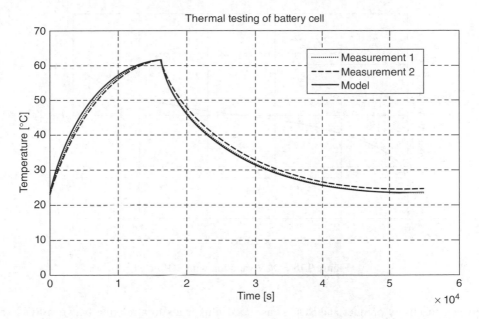

Figure 9.16 Comparison of test and calculation of the thermal response of a lithium-ion battery cell.

when discharging. The size of the reversible heating effect may be found experimentally by measuring dV_{oc}/dT, but at C-rates greater than one it is usually negligible compared with the irreversible (Joule) heating.

The cell voltage has been modelled by subtracting the voltage drop across the temperature dependent internal resistance from the OCV V_{oc}, while V_{oc} may be calculated using the Nernst equation (9.21) [20]:

$$V_{cell} = V_{oc} - IR_i(T) \tag{9.20}$$

$$V_{oc} = E_0 + \frac{RT}{nF} \ln\left(\frac{A_p}{A_r}\right) \tag{9.21}$$

where E_0 is a cell-specific constant and R is the gas constant. The variables A_p and A_r are the activity products, of the chemical products and reactants respectively. If a linear relation

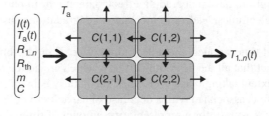

Figure 9.17 Heat exchange between cells in a two-dimensional array.

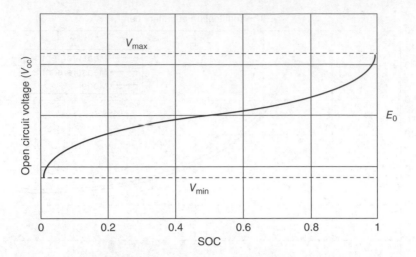

Figure 9.18   OCV variation with SOC level.

between reactivity product and SOC is assumed, this leads to a relation between SOC and OCV which is shown in Figure 9.18.

9.3.3   Modelling Degradation Effects

The thermal part of the model is integrated in the main program that consists of a time loop in which electrical properties, degradation effects and thermal properties are calculated in each time step. Figure 9.19 shows a graphical representation of the model. Besides the thermal and electrical parameters, main inputs are the charge–discharge current profile and the ambient temperature profile.

Since thermal effects, and in particular degradation effects, have rather long time scales, a relatively large time step of, for example, 5 min will provide sufficient time resolution to accurately capture the dynamic effects.

The time loop contains an equation-based description of the degradation effects, with the formerly mentioned rain-flow-counting algorithm as a central part. The rain-flow-counting calculation is rather time consuming and, therefore, it is only activated at certain time intervals.

Degradation is assumed to result in reduced battery capacity and increased internal resistance. Besides the degrading effects resulting from SOC cycling and thermal cycling, the C-rate in itself is thought to have a degrading effect. This might in part be attributed to local heating effects in the electrodes, leading to mechanical degradation caused by uneven thermal expansion.

The assumption that time, as such, has a degrading effect should probably be seen more as a recognition of the fact that all the degrading effects, which are acting together in a complex manner, are to some extent taking place even when the battery is in an idle state.

Typically, end of life is associated with a decrease in capacity to 80% of the original value. Simulating a typical use-pattern for a representative amount of time (e.g. a few months) and observing the corresponding reduction in battery capacity will enable an extrapolation to

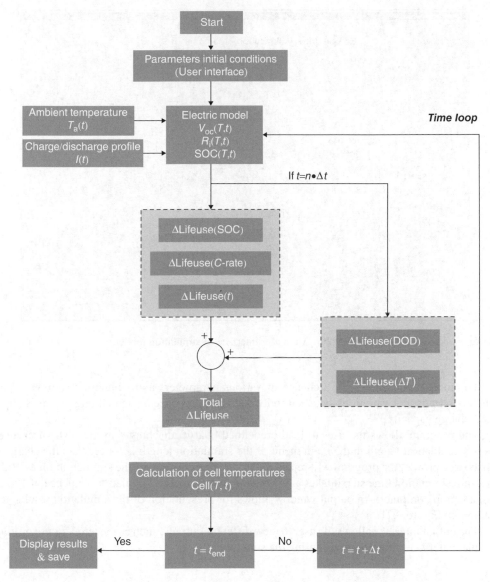

Figure 9.19 Sketch of battery degradation simulation model.

end of life and thereby establish a link between use-pattern and battery lifetime. A thorough investigation of issues related to battery modelling may also be found in Chaturvedi *et al.* [21].

Model Description and Use

The battery model has been implemented in MATLAB® and includes a graphical user interface. The main user interface is shown in Figure 9.20.

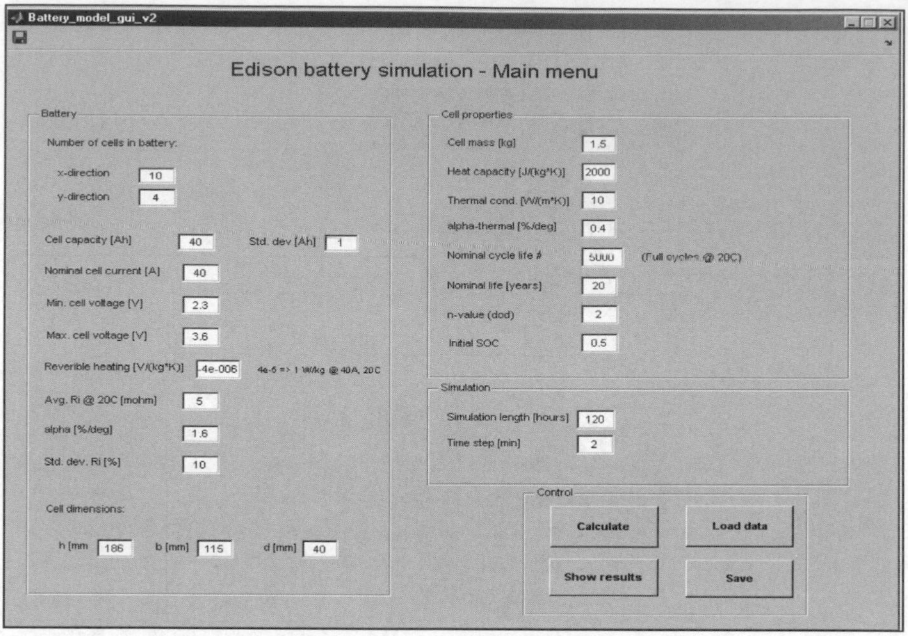

Figure 9.20   Main user interface of simulation model.

The main menu allows for a change of various parameters and simulation settings. This includes statistical variations between individual cell properties, such as internal resistance and charge capacity.

The program allows the user to load predefined charge–discharge profiles. The user also sets a simulation length in the main menu. If the simulation length is longer than the charge–discharge profile, the program will repeat the profile until the end of the simulation time. With a 2 min simulation time step it takes a few minutes to calculate a simulation sequence of 120 h on a laptop computer. An output window allows for presentation of the simulation results, as shown in Figure 9.21.

The output interface allows the user to view different outputs, such as the temperature profile of the battery, at different times during the simulation.

9.3.4   Simulation of EV Use

Various test runs have been performed in order to verify that the model results are in accordance with logic and common-sense reasoning. Figure 9.22 shows a contour plot of temperature distribution during operation, inside an array of 4×10 cells. The large time-constant and variations in ambient temperature and current results in a continuously changing internal temperature profile. Differences in internal resistance and capacity amongst the cells result in a nonsymmetrical temperature distribution across the array.

When the battery is exposed to 'normal' operation, in which discharging is only related to driving, the internal temperatures do not diverge much from ambient temperature. This is due

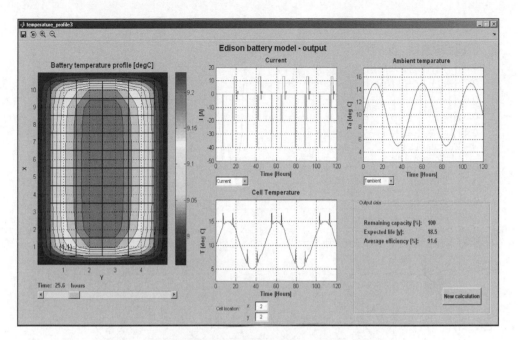

Figure 9.21   Model output interface.

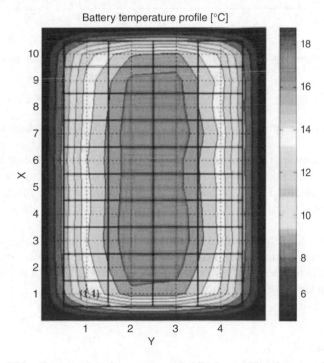

Figure 9.22   Contour plot of temperature distribution within a battery cell array.

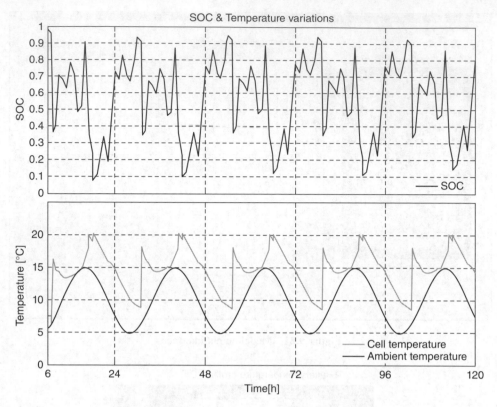

Figure 9.23 Simulated SOC and core temperature of an EV battery with combined driving and V2G operation.

to the combination of relatively low losses (in the range of 5–10%) and high thermal capacity of the cells.

If the battery is exposed to a more continuous sequence of charge and discharge, as expected when performing V2G operation, the battery core temperature tends to vary approximately 5–10 °C above ambient. This is indicated in Figure 9.23, where the result of a 120 h test run is shown. The simulated charge–discharge pattern is the result of combined driving and V2G operation.

The simulations show that such semi-continuous operation leads to an overall increase in efficiency because of a temperature increase which leads to reduced internal resistance.

The overall performance that the user experiences is determined by a number of important parameters which are closely interrelated in a complex manner:

- charge capacity
- maximum power
- efficiency
- battery degradation (lifetime).

Table 9.2 Overview of test simulation profiles.

Profile	Description
1	Work–home driving profile with relatively fast charging in the evening.
2	Work–home driving profile with slow charging during both day and night.
V2G1	The same as 2 but with V2G operation resulting in random inter-hour SOC variations of ±0.15.
V2G2	The same as V2G1 but with SOC variations of ±0.2.
V2G3	The same as V2G1 but with SOC variations of ±0.25.

The simulation model presented here enables a closer study of these relations, but this has yet to be done.

In order to obtain an indicative result with regard to the effect of use-pattern, five different charge–discharge profiles have been simulated. Each profile had a length 24 h but they were repeated seven times to reach a total simulation length of 168 h. The five profiles, which were artificially made to indicate the effect of different use-patterns, are listed in Table 9.2.

The SOC profile simulations should only be regarded as rough indicators of how such an operation might affect battery life, as well as an indication of how a thorough simulation-based analysis can contribute further. Figure 9.24 shows profiles 2 and V2G1.

The simulated battery contained 40 cells, each with a capacity of 40 Ah. The cycle life at 100% DOD was set to 5000 cycles. Results from the five test simulations are listed in Table 9.3. It shows a significant reduction in battery life as a result of V2G operation, which is not surprising since it leads to an increased number of partial cycles. The effect of slow day- and night-charging – as opposed to fast charging – is to increase the expected lifetime. This is because it lowers the overall C-rate and reduces the DOD level.

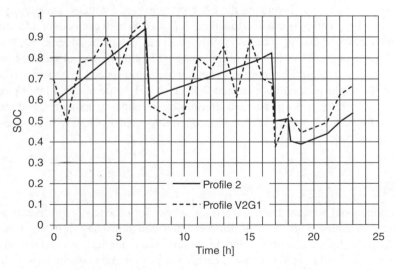

Figure 9.24 Two of the simulated SOC profiles.

Table 9.3   Results from test simulations.

Profile	1	2	V2G1	V2G2	V2G3
Expected life (years)	19.8	27	14	10.7	9.7
Efficiency (%)	90.8	94.1	91.4	94.8	95.5

The lifetime reducing effect of increased DOD level is also seen from the fact that increasing the size of SOC variations in V2G mode results in significantly shorter lifetime.

The overall efficiency of the battery is higher for profile 2 than for profile 1. This is not surprising since both DOD and C-rate during charging are lower for profile 2. But surprisingly, there is the opposite tendency for the three V2G profiles. Beyond a certain point, larger SOC cycles seem to result in higher efficiency. This behaviour can be explained by the fact that more-or-less continuous operation results in higher cell temperature and thereby lower internal resistance. These examples illustrate the complexity of this system of interrelated and sometimes counteracting mechanisms and the usefulness of a simulation model.

Thermal Management of EV Batteries

The increase of internal resistance with decreasing temperature is shown to reduce overall efficiency, but it also has the effect of reducing battery power output at low temperatures. In addition to this, the battery capacity also decreases with temperature, leading to decreased driving range. It therefore seems relevant to consider if temperature control of EV batteries can lead to better performance.

In cold regions, temperature control could be performed by insulating the battery and installing controlled circulation and exchange of air with the surroundings. According to Al Hallaj et al. [12], reducing lithium-ion battery temperature from $+20\,^{\circ}C$ to $-10\,^{\circ}C$ leads to a reduction of battery capacity in the range of 40%. For a 20 kWh battery this corresponds to a capacity loss of 8 kWh. Insulating the EV battery package with expanded polystyrene in a thickness of a few centimetres will allow for a ΔT of 20 K between battery and ambient at a heat loss in the range of 200 W. This loss would, to a large extent, be covered by battery losses which constitutes about 70 W for slow night charging and up to several hundred watts when performing V2G operation. The remaining heat, which would need to be supplied to sustain a ΔT of 20 K, would certainly not exceed 2 kWh/day. This points to a large potential for improvement of battery performance and lifetime by introducing insulation and temperature control in relation to EV batteries, at least when operating them in places with large temperature variations.

Conclusions

The simulation model presented here has so far not been subjected to a representative selection of test cases. In addition to that, the implemented equations describing cell degradation mechanisms are so far based entirely on data obtained from publications and not on empirical data from a specific cell type. It is thus too early to draw any exact conclusions with regard to the influence that use-pattern has on lifetime.

The main purpose of this work has been to make and test a simulation model which will enable a closer study that can lead to exact conclusions with regard to the effect that use-pattern has on battery lifetime and performance. In that sense, this work has been successful.

The model will clearly be able to assist in at least two important areas: one is to complement experimental results to obtain a clearer picture of how the different mechanisms influence battery aging and the second is to determine how battery use-pattern affects aging.

The initial model studies indicate a 50% reduction in battery life as a result of V2G operation. A more detailed study might show how to prolong battery life and improve efficiency. Changing the model equations to include measurement-based inputs is a manageable task which can be done gradually as more knowledge is gained.

9.4   Electric EC Model

9.4.1   Battery Modelling: Dynamic Performance

Section 9.3 on thermal modelling and degradation details the battery characteristics in terms of thermal behaviour and driving pattern. In this section, the SOC modelling of the battery is detailed and a parameterization procedure is described. The battery models have been implemented in MATLAB®/Simulink and they represent the different chemistries of batteries used under the circumstances described below.

In the following, modelling of dynamic performance of a battery will be described, including different battery models suggested in the literature, construction of a complete battery model and parameterization of the battery model.

The battery electric characteristics, longevity and runtime are affected by the type of chemistry and ambient conditions, as well as driving and charging patterns. The modelling work basically includes two elements as indicators of battery status: SOC and SOH. The SOC identifies the available capacity of the battery, while SOH indicates the battery performance status, including actual usable capacity and remaining runtime.

Several different chemistries are used for battery applications. For the work presented here, LFP and NMC have been used. It is expected that lithium-ion chemistries will become the primary battery technology for EVs as this type of battery offers the highest energy density for EV use [22].

The voltage–current charge and discharge curve describes the electrical behaviour of a battery and it is affected by temperature, charge/discharge rate, charge/discharge cycle number, storage time and SOC. All of these factors should be considered in the battery model. Charge and discharge cannot necessarily be modelled by the same model. This depends on the type of battery and the hysteresis of the battery. The battery impedance is also modelled by the battery model. As the V–I characteristic will be modelled as described above, the impedance will also be known. Aging is also represented in the battery model. It occurs through thermal, cyclic and storage aging.

Power losses and self-discharge are also important. The self-discharge is not represented in the battery model presented here but can be found from the battery current and impedances. According to Johnson and White [23], a commercially available lithium-ion battery cell shows a self-discharge of 3% of the initial capacity over 30 days. Self-discharge effects, therefore, are of minor significance for applications where the battery is used often. An example of a self-discharge curve is shown in Figure 9.25.

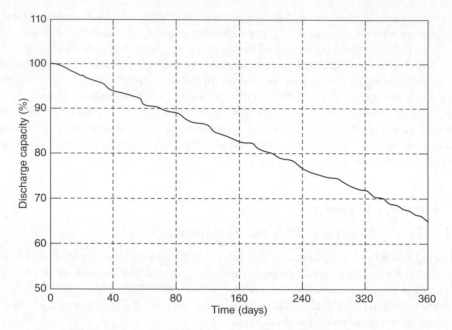

Figure 9.25   Example of a battery self-discharge curve at 20 °C.

9.4.2   *Battery Cell Models Described in the Literature*

The most promising battery systems for EVs and plug-in hybrid EVs are based on lithium-ion or lithium-polymer batteries, which we will focus on here. A variety of battery models have been proposed in the literature to capture the battery characteristics from different perspectives. This model classification is inspired by Chen and Rincon-Mora [24] and different classifications are used in Rao *et al.* [25] which give a good overview of different battery models. Roughly, these models can be classified into electrochemical [21, 26, 27], mathematical [28] and equivalent electric circuit models [24, 29, 30] to serve different study purposes.

For grid impact studies, the battery model must be kept as simple as possible in order to facilitate the simulation of a large number of EVs into the grid; while from the owners it is desirable to know the current battery SOC and SOH, as well as the battery runtime prediction [31]. Electric circuit models are seen as an intermediate approach [29] between electrochemical and mathematical models where the model accuracy – in the range of 1 to 5% [24] – and computational efficiency are balanced. In Gao *et al.* [29], electrical and thermal properties are considered in the model formulation, where the EC consists of an *RC* circuit with a series resistance. The work of Chen and Rincon-Mora [24] extends the work of Gao *et al.* [29] with two *RC* circuits representing the short- and long-term transients, and the battery runtime is also addressed in the model, though without considering the temperature and cycle number dependency. Erdinc *et al.* [30] include an additional resistance to represent the internal resistance variation with respect to the increase of cycle number, thereby taking into account the capacity fading due to time, temperature and cycle number. We do not intent to describe the many different models here, only to give a short description of the different classes.

Physical/Electro-Chemical Models

Electrochemical models include the internal workings of the battery. Thus, the physics of the battery is included in electrochemical models, which makes them very demanding with regard to computation power. Electrochemical models are typically used to optimize the battery composition, as the physical design aspects are represented in the model. These types of models, therefore, are not seen as a viable option for driving-pattern simulations or long-term degradation modelling.

Mathematical/Empirical Models

According to Chen and Rincon-Mora [24], mathematical battery models are generally very abstract with very little practical meaning. They can be used for predicting battery runtime, efficiency and/or capacity, but they are often developed for specific applications. Mathematical battery models, therefore, are also not an option for driving-pattern simulations.

Electrical-Circuit Models

Electrical-circuit models are based on representation of the battery by an electrical circuit. This often utilizes a controlled voltage source that varies the voltage with the SOC of the battery. The voltage source is connected in series with an internal resistance and an *RC* circuit which is used to model the dynamic voltage response of the battery. Depending on the layout of the electrical circuit, it can be classified as Thevenin based, impedance based, runtime based or as a mixed model. For more information, consult Chen and Rincon-Mora [24] and Araujo Leão *et al.* [22].

A battery cell model is included in the MATLAB®/Simulink SimPowerSystem package. This model is based on theory from Tremblay *et al.* [32] and consists of an internal resistance in series with a controlled voltage source. The model is depicted in Figure 9.26.

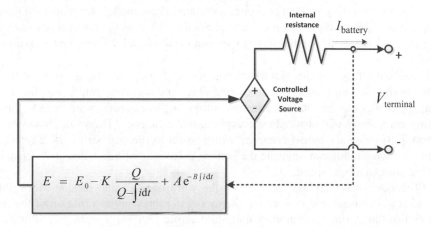

Figure 9.26 Sketch of nonlinear battery model.

The internal resistance in this model is a constant, either supplied directly from the manufacturer or it can be evaluated from the battery efficiency, nominal voltage and nominal capacity with a reasonable precision. The controlled voltage source represents the SOC-dependent no-load voltage of the battery. The expression for the no-load voltage E is

$$E = E_0 - K \frac{Q}{Q - \int_0^t i \, dt} + A e^{-B \int_0^t i \, dt} \qquad (9.22)$$

The no-load voltage (or OCV) is dependent on the battery constant voltage E_0, the polarization voltage K, the battery capacity Q (Ah), the exponential part amplitude A (V), the exponential part time constant inverse B (Ah^{-1}), time t, and the current i through the battery.

Parameters for several different battery chemistries and different cell sizes, as well as an explanation of the parameterization, are included in Tremblay et al. [32]. In general, the charge and discharge curves for the battery cell, as well as basic information on cell capacity, should be sufficient for parameterization. The model implemented in MATLAB® takes into account several different chemistries and battery sizes and it thus represents a fast and easy way to create a fairly accurate battery model.

The capabilities of the model are limited to describing the charge and discharge curves of a battery cell at constant temperature and aging. The model can simulate the transient state of a battery cell and the level of accuracy will be dependent on the information available from the manufacturer. The electrical-circuit model in MATLAB® seems simplified, compared with other electrical-circuit models, but this is also a benefit because it enables simulation with only a few input parameters.

9.4.3   Battery Model Implementation in MATLAB®

The most commonly suggested battery models for investigation of the performance of electrical vehicles are based on (mixed) electrical-circuit models. These models are more advanced than the simple electrical-circuit model presented in Section 9.4.2, as the circuit is expanded with one to three resistances in parallel with a capacitance to model the transient response more accurately.

Given the advantages of this electric-circuit model type, it has been used for the work presented here to obtain a combination of simplicity and accuracy. Intuitively, the accuracy of a battery model can be enhanced in proportion to the complexity of the EC; however, increasing complexity will also reduce computational efficiency. The work shown in Zhang and Chow [33] examines model accuracy with respect to the number of RC circuits and it is found that two circuits can replicate the battery behaviour with less than 10% error, while preserving computational speed.

The EC model presented here is based on the work of Chen and Rincon-Mora [24]. An additional series resistance is included in the model to represent the variation of the internal resistance resulting from cycle number and temperature [30, 34]. Figure 9.27 illustrates the proposed equivalent circuit.

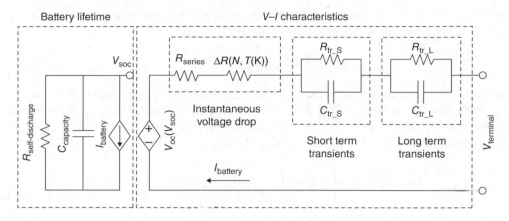

Figure 9.27   The EC model proposed in this project.

The left side of the model represents the battery SOC level, while the right side describes the battery *V–I* characteristics during charging/discharging. $C_{capacity}$ represents the present capacity of the battery. The various components of the model are as follows:

R_{series}	describes the instantaneous terminal voltage of the step current under normal temperature and low cycle number;
$\Delta R(N, T(K))$	describes the change of R_{series} due to cycle number N and ambient temperature $T(K)$;
$R_{tr\_S}, C_{tr\_S}, R_{tr\_L}, C_{tr\_L}$	resistors and capacitors of two RC circuits, describing the behaviour of battery subject to step current;
$V_{OC}(V_{SOC})$	OCV as a function of SOC;
$C_{capacity}$	usable capacity of the battery, mainly dependent on the cycle number and temperature;
$R_{self-discharge}$	battery discharging effect with respect to calendar time.

The terminal voltage can be calculated by

$$V_{terminal} = V_{OC}(V_{SOC}) - i_{battery} Z_{eq}(SOC) \qquad (9.23)$$

where Z_{eq} is the equivalent impedance of the battery, which is a function of the SOC. SOC can be calculated using

$$SOC = SOC_{init} - \int (i_{battery}/C_{capacity}) \, dt \qquad (9.24)$$

where the SOC_{init} is the initial state of charge and $i_{battery}$ is the battery current.

The internal impedance is a function of temperature, cycle number and SOC. This influences both the steady-state and dynamic characteristics of the battery. Here, we ignore the influences of temperature and cycle number on the dynamic characteristics of the battery. V_{OC} may be expressed by the following equation [29]:

$$V_{OC} = C_1 + C_2 \times SOC + C_3 \times SOC^2 + \cdots + C_n \times SOC^{n-1} \qquad (9.25)$$

where $C_1, C_2, \ldots, C_n$ are coefficients that need to be determined empirically through realistic tests. For SOC-dependent impedances, the general expression is

$$Z = a_1 + a_2 e^{SOC \times a_3} \tag{9.26}$$

where Z is the impedance of an internal resistance in series with two RC circuits. For every Z, the coefficients a_1, a_2 and a_3 will be different and measurement data are required to determine the parameters. $\Delta R(N, T(K))$ is a correction to the internal impedance which is identified from empirical data. The total impedance of the battery is

$$Z_{eq} = R_{series} + R(N, T) + R_{tr\_S}\left[1 - \exp\left(-\frac{t}{R_{tr\_S}C_{tr\_S}}\right)\right] + R_{tr\_L}\left[1 - \exp\left(-\frac{t}{R_{tr\_L}C_{tr\_L}}\right)\right] \tag{9.27}$$

One of the key issues of the battery model is an expression of available battery capacity. The capacity loss is normally considered as coming from two parts: calendar life loss and battery cycle life loss. The calendar life loss is time and temperature dependent, while the cycle life loss is related to the cycle number, depth of discharge of the individual cycles and temperature [35]. Ramadass *et al.* [28] gave mathematical expressions for the calendar life loss and cycle life loss. The calendar life loss can roughly be expressed as

$$\text{Calendar}_{\text{life loss}} = k_1 \exp\left[-\frac{k_2}{T}\right] t^{k_3} \tag{9.28}$$

where k_1, k_2 and k_3 are coefficients and T (K) is temperature. For EV batteries the cycle life loss is considered more important than the calendar life loss. In the literature, cycle life loss is normally modelled as a physical process [28] or from empirical data [15]. In this work, an empirical method has been used:

$$\text{Cycle}_{\text{life loss}} = f(N, T) \tag{9.29}$$

The available capacity of the battery can be expressed as

$$C_{available} = C_{init}(1 - \text{Calendar}_{\text{life loss}} - \text{Cycle}_{\text{life loss}}) \tag{9.30}$$

where C_{init} is the initial capacity of the battery. Battery end of life is usually associated with the capacity being equal to or below 80% of the initial capacity.

Model Construction

A simulation model of a lithium battery, based on the analytical model description of the previous section, was built in MATLAB®/Simulink. The inputs and outputs of the model are listed in Table 9.4 and an example of charge–discharge simulation curves is shown in Figure 9.28. The model was created in Simulink and the basic structure is shown in Figure 9.29.

Table 9.4 Model inputs and outputs.

Inputs	Outputs
Initial SOC	Battery terminal voltage (V)
Initial battery capacity (Ah)	SOC
Temperature (K)	
No. of cycles used N	
Time used (months)	
Cut-off voltage (V)	
Charging–discharging current (A)	

From Battery Cell Model to Battery Pack Model

The battery model described here was intended for modelling a battery cell, not an entire battery pack. The question then is how the full battery pack should be modelled. The pack model can either be constructed by combining an appropriate number of cells in series (to increase the voltage across the pack) or in parallel (to increase the capacity of the pack). The large number of battery parameters will significantly increase the computational burden.

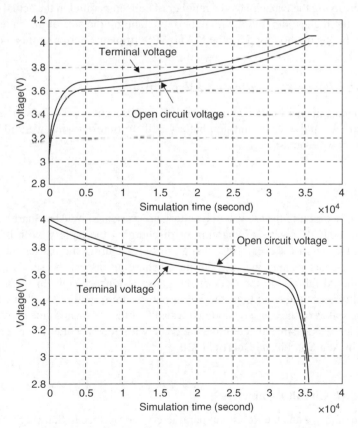

Figure 9.28 Example of charge and discharge curves.

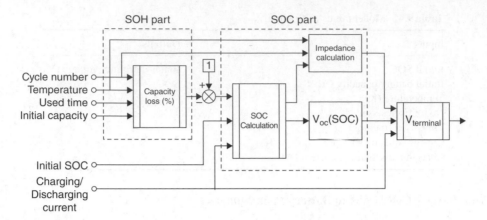

Figure 9.29   Battery model structure.

Another approach would be to scale the cell model to a pack model. This would include multiplying the voltage of the cell model with the number of cells in series in the actual battery as well as multiplying the capacity by the number of cells in parallel in the actual battery. This method offers the highest flexibility with regard to changing the size of the battery pack and, therefore, will be the method used for the MATLAB® implementation of this model.

9.4.4   Model Parameterization and Validation

This section details the parameterization and validation procedures. The parameters presented here are based on a polymer lithium-ion battery cell. In the following we will present how these parameters were obtained.

Extraction of OCV

Two methods of extracting the OCV of the battery are presented by Abu-Sharkh and Doerffel [32]. The first method consists of charging or discharging the battery, with built-in pauses of approximately 1 min for every 10% change in SOC. During the pauses the voltage will decrease during charging and increase during discharging, as can be seen in Figure 9.30. The voltage should fall/rise to the OCV if the pause is long enough. In Figure 9.30 it can be seen that this is not the case; thus, the 1 min pauses are in reality too short. However, by connecting the spikes of the pause by dotted lines and taking the mean value between charging and discharging, it is possible to determine the OCV with good accuracy. Further information can be found in Abu-Sharkh and Doerffel [32].

Extraction of *RC* Circuit Parameters

The extraction of R_{series} and the *RC* circuit parameters is described in Schweighofer *et al.* [36]. The battery is subjected to charge and discharge current pulses of different values (0.5–200 A)

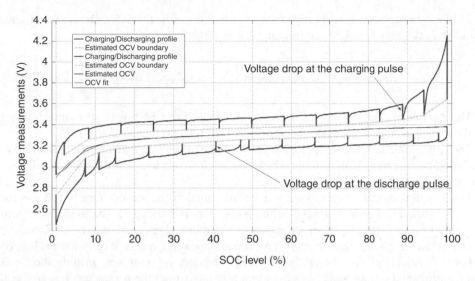

Figure 9.30 Graphical explanation of how to determine OCV from charge and discharge curves.

for 5 s. As the *RC* circuit represents the short- and long-term transient responses respectively, the parameters can be found through analysing the step response.

The short-term response parameters can be found by looking at the first 0.5 s of the response. This also applies for the series resistance, which is responsible for the instantaneous voltage rise when the current pulse is applied. The instantaneous voltage rise U_R is marked in Figure 9.31. The series resistance can then be found by dividing U_R by the amplitude of the current pulse.

A time constant τ_D can be read from the initial slope of the voltage response. The short transient response can be described by the following equation:

$$u(t) = U_R + \hat{U}_D[1 - \exp(-t/\tau_D)] \tag{9.31}$$

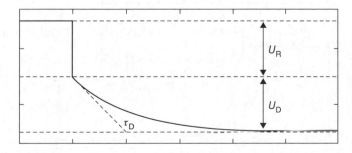

Figure 9.31 Determining the electrical circuit parameters from the voltage response to a current pulse.

The long-term response can be neglected when looking at the first 0.5 s. The electrical-circuit parameters can now be calculated from the following three equations:

$$R_{series} = \frac{U_R}{I} \qquad R_{transientS} = \frac{\hat{U}_D}{I} \qquad C_{transientS} = \frac{\tau_D}{R_{transientS}} \qquad (9.32)$$

The parameters for the long-term transient response can be found in a similar way by analysing the voltage response in the remaining time interval (from 0.5 to 5 s).

The expression and parameters for the calendar life losses originate from Spotnitz [37] and are based on data from Toshiba lithium batteries, but it is not clear how the calendar life losses are parameterized.

The Simulink model developed needs to be confirmed by the manufacturer's data sheet and by the experimental results. Tuning of the parameters is basically a nonlinear curve-fitting problem. Owing to the large number and large variation range of parameters involved in the model, it may not be efficient to apply a nonlinear optimization tool directly for this purpose.

For the model deployed, the validation needs to be done by combining manufacturer's data and experimental results, such as step-current tests, to extract the parameters that reflect the steady-state, dynamic and runtime characteristics of the battery. However, based on the model structure, it is observed that the model can almost be decoupled into an SOC part and an SOH part, where each of them may be validated separately based on the manufacturer's data sheet and test results. In this work, a two-step validation procedure is proposed to validate the whole battery model.

The validation of the SOC part involves the parameters of V_{OC} and the impedance parameters. In the first step, the parameters of the battery are validated under normal temperature with a very low cycle number, and the value of $\Delta R(N, T)$ is assumed to be zero. A step-current test is required to extract the parameters of the battery, and the discharging curve from the manufacturer's data sheet can be used to confirm the steady-state characteristics model. Figure 9.32 shows the model for step-current testing. The final impedance may be confirmed with the EIS test.

It is recommended that SOH validation is done together with a test of $\Delta R(N, T)$. This test can also be decoupled into two tests. In the first test, the cycle number is fixed while the battery is tested under different ambient temperatures. In the second test, the cycle number will be varied under constant temperature. The discharge curves provided by the manufacturer for different

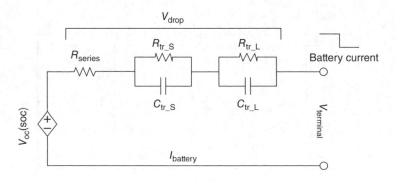

Figure 9.32   Step-current test.

temperatures and cycle numbers can be used as reference for the discharging characteristics as well as for analysing the variation of $\Delta R(N, T)$ with respect to cycle number and temperature.

Future Work

The dynamic characteristics of a battery are functions of SOC, cycle number and temperature [38]. In this chapter we ignore the variation in dynamic characteristics of the battery with temperature and cycle number while mainly focusing on the steady-state characteristics. The dynamic part can be further included in the validation work by performing step-current tests at different temperatures and cycle numbers to make an empirical analysis of the battery characteristics. Also, for grid integration, the EV battery model should be at battery-pack level instead of cell level. However, a solid cell model is essential in order to form a good basis for modelling pack performance. The development of a pack model should mainly be based on test results and real EV charging characteristics.

References

[1] Safari, M., Morcrette, M., Delacourt, C. and Teyssot, A. (2010) Life prediction methods for lithium-ion batteries derived from a fatigue approach. II. Capacity-loss prediction of batteries subjected to complex current profiles. *Journal of the Electrochemical Society*, **157**, A892–A898.

[2] Tröltzsch, U., Kanoun, O. and Tränkler, H.R. (2006) Characterizing aging effects of lithium ion batteries by impedance spectroscopy. *Electrochimica Acta*, **51**, 1664–1672.

[3] Vetter, J., Novák, P., Wagner, M.R. *et al.* (2005) Ageing mechanisms in lithium-ion batteries. *Journal of Power Sources*, **147**, 269–281.

[4] Barsoukov, E., Ryu, S.H. and Lee, H. (2002) A novel impedance spectrometer based on carrier function Laplace-transform of the response to arbitrary excitation. *Journal of Electroanalytical Chemistry*, **536**, 109–122.

[5] Van Ingelgem, Y., Tourwé, E., Vereecken, J. and Hubin, A. (2008) Application of multisine impedance spectroscopy, FE-AES and FE-SEM to study the early stages of copper corrosion. *Electrochimica Acta*, **53**, 7523–7530.

[6] Van Ingelgem, Y., Tourwé, E., Blajiev, O. *et al.* (2009) Advantages of odd random phase multisine electrochemical impedance measurements. *Electroanalysis*, **21**, 730–739.

[7] Boukamp, B.A. (2004) Electrochemical impedance spectroscopy in solid state ionics: recent advances. *Solid State Ionics*, **169**, 65–73.

[8] Onda, K., Nakayama, M., Fukuda, K. *et al.* (2006) Cell impedance measurement by Laplace transformation of charge or discharge current–voltage. *Journal of the Electrochemical Society (USA)*, **153**, A1012–A1018.

[9] Klotz, D., Schönleber, M., Schmidt, J.P. and Ivers-Tiffée, E. (2011) New approach for the calculation of impedance spectra out of time domain data. *Electrochimica Acta*, **56**, 8763–8769.

[10] Abu-Sharkh, S. and Doerffel, D. (2004) Rapid test and non-linear model characterisation of solid-state lithium-ion batteries. *Journal of Power Sources*, **130**, 266–274.

[11] Williford, R.E., Viswanathan, V.V. and Zhang, J.G. (2009) Effects of entropy changes in anodes and cathodes on the thermal behavior of lithium ion batteries. *Journal of Power Sources*, **189**, 101–107.

[12] Al Hallaj, S., Venkatachalapathy, R., Prakash, J. and Selman, J.R. (2000) Entropy changes due to structural transformation in the graphite anode and phase change of the $LiCoO_2$ cathode. *Journal of the Electrochemical Society*, **147**, 2432–2436.

[13] Al Hallaj, S., Prakash, J. and Selman, J.R. (2000) Characterization of commercial Li-ion batteries using electrochemical–calorimetric measurements. *Journal of Power Sources*, **87**, 186–194.

[14] Sato, N. (2001) Thermal behavior analysis of lithium-ion batteries for electric and hybrid vehicles. *Journal of Power Sources*, **99**, 70–77.

[15] Kempton, W., Tomic, J., Letendre, S. *et al.* (2001) Vehicle-to-grid power: battery, hybrid, and fuel cell vehicles as resources for distributed electric power in California, http://www.udel.edu/V2G/docs/V2G-Cal-2001.pdf (accessed January 2013).

[16] Marra, F., Træholt, C., Larsen, E. and Wu, Q. (2010) Average behavior of battery-electric vehicles for distributed energy studies, in *2010 IEEE PES Innovative Smart Grid Technologies Conference Europe (ISGT Europe)*, IEEE, Piscataway, NJ.

[17] Sauer, D.U. and Wenzl, H.W. (2007) Comparison of different approaches for lifetime prediction of electrochemical systems – using lead–acid batteries as example. *Journal of Power Sources*, **176**, 534–546.

[18] Downing, S.D. and Socie, D.F. (1982) Simple rain-flow counting algorithms. *International Journal of Fatigue*, **4** (1), 31–40.

[19] Winter, M. and Brodd, R.J. (2004) What are batteries, fuel cells, and super capacitors? *Chemical Reviews*, **104** (10), 4245–4270.

[20] Chaturvedi, N.A., Klein, R., Christensen, J. *et al.* (2010) Algorithms for advanced battery-management systems – modeling, estimation and control challenges for lithium-ion batteries. *IEEE Control Systems Magazine*, **30** (3), 49–68.

[21] Schmidt, A.P., Bitzer, M., Imre, R.W. and Guzzella, L. (2010) Experiment driven electrochemical modelling and systematic parameterization for a lithium-ion battery cell. *Journal of Power Sources*, **195** (15), 5071–5080.

[22] Araujo Leão, J.F., Hartmann, L.V., Correa, M.B.R. and Lima, A.M.N. (2010) Lead–acid battery modeling and state of charge monitoring, in *Twenty-Fifth Annual IEEE Applied Power Electronics Conference and Exposition (APEC)*, IEEE, Piscataway, NJ, pp. 239–243.

[23] Johnson, B.A. and White, R.E. (1998) Characterization of commercially available lithium-ion batteries. *Journal of Power Sources*, **70** (1), 48–54.

[24] Chen, M. and Rincon-Mora, G.A. (2006) Accurate electrical battery model capable of predicting runtime and I–V performance. *IEEE Transactions on Energy Conversion*, **21** (2), 504–511.

[25] Rao, R., Vrudhula, S. and Rakhmatov, D.N. (2003) Battery modeling for energy aware system design. *Computer*, **36** (12), 77–87.

[26] Bashash, S., Moura, S.J., Forman, J.C., and Fathy, H.K. (2011) Plug-in hybrid electric vehicle charge pattern optimization for energy cost and battery longevity. *Journal of Power Sources*, **196** (1), 541–549.

[27] Urbain, M., Hinaje, M., Rael, S. *et al.* (2010) Energetical modeling of lithium-ion batteries including electrode porosity effects. *IEEE Transactions on Energy Conversion*, **25** (3), 862–872.

[28] Ramadass, P., Haran, B., White, R. and Popov, B.N. (2003) Mathematical modeling of the capacity fade of Li-ion cells. *Journal of Power Sources*, **123** (2), 230–240.

[29] Gao, L., Liu, S. and Dougal, R.A. (2002) Dynamic lithium-ion battery model for system simulation. *IEEE Transactions on Components and Packaging Technologies*, **25** (3), 495–505.

[30] Erdinc, O., Vural, B. and Uzunoglu, M. (2009) A dynamic lithium-ion battery model considering the effects of temperature and capacity fading, in *2009 International Conference on Clean Electrical Power (ICCEP 2009)*, IEEE, Piscataway, NJ, pp. 383–386.

[31] Binding, C., Gantenbein, D., Jansen, B. *et al.* (2010) Electric vehicle fleet integration in the Danish Edison project – a virtual power plant on the island of Bornholm, in *2010 IEEE Power and Energy Society General Meeting*, IEEE, Piscataway, NJ.

[32] Tremblay, O., Dessaint, L.-A. and Dekkiche, A.-I. (2007) A generic battery model for the dynamic simulation of hybrid electric vehicles, in *IEEE Vehicle Power and Propulsion Conference, 2007. VPPC 2007*, IEEE, Piscataway, NJ, pp. 284–289.

[33] Zhang, H. and Chow, M.-Y. (2010) Comprehensive dynamic battery modeling for PHEV applications, in *2010 IEEE Power and Energy Society General Meeting*, IEEE, Piscataway, NJ.

[34] Zhang, S., Xu, K. and Jow, T. (2004) Electrochemical impedance study on the low temperature of Li-ion batteries. *Electrochimica Acta*, **49** (7), 1057–1061.

[35] Dai, H., Wei, X. and Sun, Z. (2009) A new SOH prediction concept for the power lithium-ion battery used on HEVs, in *IEEE Vehicle Power and Propulsion Conference, 2009. VPPC '09*, IEEE, Piscataway, NJ, pp. 1649–1653.

[36] Schweighofer, B., Raab, K.M. and Brasseur, G. (2003) Modeling of high power automotive batteries by the use of an automated test system. *IEEE Transactions on Instrumentation and Measurement*, **52** (4), 1087–1091.

[37] Spotnitz, R. (2003) Simulation of capacity fade in lithium-ion batteries. *Journal of Power Sources*, **113** (1), 72–80.

[38] Bloom, I., Cole, B., Sohn, J. *et al.* (2001) An accelerated calendar and cycle life study of Li-ion cells. *Journal of Power Sources*, **101** (2), 238–247.

10

Day-Ahead Grid Tariffs for Congestion Management from EVs

Niamh O'Connell,[1] Qiuwei Wu[2] and Jacob Østergaard[2]

[1] Department of Applied Mathematics and Computer Science, Technical University of Denmark, Kgs. Lyngby, Denmark

[2] Centre for Electric Technology, Department of Electrical Engineering, Technical University of Denmark, Kgs. Lyngby, Denmark

10.1 Introduction

As generation portfolios develop towards ever-increasing shares of renewable generation, system operators are challenged to maintain system stability and security in the face of fluctuating generation. This level of variability is likely to necessitate increased generation reserves to ensure that contingencies such as wind ramping events can be survived. Alternatively, or perhaps in combination with increased reserves, demand flexibility can be harnessed to bring the system from the paradigm in which generation follows demand to one in which flexible demand follows the generation pattern, thus facilitating higher penetrations of variable generation sources such as wind or solar power. Significant research is currently ongoing to determine the flexibility capabilities of conventional demand; however, the inherent flexibility of electric vehicle (EV) charging promises significant scope for the provision of system services. EVs are typically available for charging for long periods of time (Kempton and Tomić, 2005) and often are not charging for the entire plugged-in period. This enables the provision of peak-shaving and generation-following through intelligent scheduling of charging, and possibly even the provision of ancillary services such as regulating power through power injection from the EV, known as the vehicle-to-grid (V2G) concept (Kempton and Tomić, 2005).

Provision of any meaningful system contribution will require a large number of EVs; while this will naturally result in increased system demand requiring increased system capacity, another, perhaps more pressing issue, is their impact on the distribution system. Legacy grids that were designed for residential-scale loading will be faced with additional demands on a much larger scale, which without management will necessitate widespread grid reinforcement and upgrade.

10.1.1 Power System Congestion

Power system congestion is defined as 'a consequence of network constraints characterizing a finite network capacity that precludes the simultaneous delivery of power from an associated set of power transactions' (Singh et al., 1998). Conventional grid congestion occurs mainly on meshed networks at higher voltage levels and presents in the form of transmission line overload. Vertically integrated power systems manage this through generation rescheduling; this results in increased system costs, which in turn drives investments for network upgrade and reinforcement. This approach is sufficient when a single utility controls the system; however, in a deregulated power system, this responsibility is transferred to the transmission system operator (TSO). The TSO is charged with ensuring security of supply, thus ensuring that transmission lines remain in service and are not overloaded. A number of approaches to transmission congestion management have been adopted; these vary from TSO-determined generation rescheduling with associated compensation (such as in the UK), to nodal pricing (such as in PJM) (Singh et al., 1998; Hamoud and Bradley, 2004), and zonal pricing (such as on the Nordic market) (Kumar et al., 2004).

EV demands are expected to occur primarily at the low voltage (LV) distribution level; as such, the congestion will be concentrated at the LV and medium voltage (MV) levels. Many studies have been conducted analyzing congestion issues on the MV network; however, they note that the problems likely originate on the LV network; as such, analysis of this network should be conducted as a first step in congestion studies (Lopes et al., 2009; Maitra et al., 2010; Taylor et al., 2009; Zhao et al., 2010). It is noted by Sundstrom and Binding (2011) that EV charging will have a major impact on distribution system operators (DSOs). As such, they will be incentivized to support strategically scheduled EV charging if it means that massive network upgrades can be avoided. The congestion management procedures mentioned previously are not designed for use at lower voltage levels; thus, alternative approaches must be devised.

Grid congestion resulting from EV demand manifests in many different forms, depending on the penetration of EVs, their distribution on the network, the available capacity on the network, and the local grid topology. Grid congestion occurs most commonly in the form of transformer overload, with line loading also an issue in some cases. A study on the impact of uncontrolled EV charging in British Columbia (Kelly et al., 2009) found that suburban networks are most likely to experience transformer overload due to high customer-to-transformer ratios, while rural networks are more susceptible to voltage drop due to longer feeder lengths.

It is generally assumed that a certain penetration of EVs is tolerable on most networks, even assuming uncontrolled charging; Lopes et al. (2009) detailed a study of EV penetration on an MV distribution network and noted that a penetration of 10% EVs with uncontrolled charging is possible without causing grid congestion. That study, however, did not specify the distribution of EVs analyzed. Another study (Taylor et al., 2010) highlighted the possible impacts of clustered EVs. It is noted that clustered EVs result in lower spatial diversity and, consequently, increased probability that components close to the EV load will experience congestion. Most notably, that study concluded that, given the possibility of clusters, negative impacts of EVs cannot be disregarded even for penetrations as low as 2%. This view is supported by a study on Hydro Quebec's distribution system (Maitra *et al.*, 2010) which expressed concern for the impact of EV clusters even in the early stages of consumer adoption. However, stochastic analysis by Taylor *et al.* (2010) found that, generally, spatial and temporal diversity of charging prevents excessive congestion for penetrations up to 8%.

Gong *et al.* (2011) conducted a study on the impact of EV charging on distribution transformer lifetimes. A detailed thermal model of the transformer was used to determine the aging acceleration due to hotspots in the transformer winding. The hotspot temperature was modeled as a function of the current loading and load losses. The study concluded that uncontrolled charging at peak periods would cause intolerable insulation-aging acceleration to such an extent that loading-based failure would occur prior to insulation-based failure. This worst-case scenario was based on simultaneous charging of six EVs on a local distribution network; randomized charging patterns were adopted as a 'low-tech' approach to coordinated charging. In this case, the impact of insulation aging was reduced, but it still limits the overall transformer lifetime.

10.1.2   Coordinated EV Charging

The general consensus in the existing literature on this topic is that uncontrolled charging is tolerable in most cases, from a distribution-level congestion perspective, for penetrations of up to approximately 10%. Above this level, however, controlled charging is required to mitigate the negative impacts of increased network loading.

A number of approaches to controlled charging are proposed in the literature, depending on the grid impact to be mitigated. A common approach to voltage drop is the inclusion of a grid interface device in the EV that stops charging when the voltage drops below a given threshold (Lopes *et al.*, 2011), or power electronics to provide reactive power compensation (Dyke *et al.*, 2010). These could be particularly useful on rural networks where voltage drop is the primary issue with EV charging due to longer feeder lengths. However, as noted by Lopes *et al.* (2011), this does not solve the issues related to transformer and line overload; these require charging control on a higher level.

Coordinated control of EVs is a common solution proposed in the literature; the flexible nature of EV demands facilitates the shifting of charging to avoid peak congestion periods. Coordination of charging can be either centralized or decentralized. Centralized control typically occurs under a single responsible party such as a fleet operator (FO), which ensures that optimal fleet charging occurs with respect to any number of objectives. An FO exercising centralized control for congestion management will typically be privy to the operating status of the grid and will diversify charging such that the grid constraints are respected. Decentralized control involves the transmission of a control signal to individual EVs to elicit a particular response. This often involves the broadcasting of a price signal, or set thereof, according to which each EV can individually optimize their charging behavior towards their individual objectives; for example, cost minimization.

Dual or time-of-use (TOU) tariffs are a form of indirect control suggested in a number of studies, with the Pacific Northwest National Laboratory (PNNL) conducting an in-depth study into their use for congestion alleviation. The study was primarily concerned with heating and cooling loads; however, the methodology can also be applied to EVs. They found that dual tariffs were effective at reducing the average system peak; however, there were issues relating to the switch between high- and low-tariff periods: as switch time was uniform throughout the system, there was a pronounced change in demand at that time. Also, if the peak periods were incorrectly predicted and, therefore, the tariffs applied incorrectly, the system peak could be more severe. This view is supported by Shao *et al.* (2010) and Taylor *et al.* (2010), who noted that incorrect application of control schemes can exacerbate, rather than alleviate, congestion problems. The PNNL study found that real-time pricing is most effective at alleviating congestion and resulted in the smoothest demand response (Hammerstrom *et al.*, 2007).

Decentralized charging control often occurs at, or close to, real time to facilitate the provision of an accurate representation of the current state of the grid. Real-time operations will become increasingly important as the power system evolves to include distributed generation, whose outputs can be difficult to predict but can be accurately measured in real time.

A novel approach to decentralized charging of plug-in hybrid electric vehicles (PHEVs) was proposed by Fan (2011) where grid congestion is prevented on a real-time scale through the application of the internet congestion pricing principle. Parallels are drawn between internet traffic and electrical grid congestion, as both use a single physical infrastructure for the supply of information or power, through a fixed bandwidth or capacity. The concept proposed here is based on the proportionally fair-play pricing scheme as proposed by Kelly *et al.* (1998) whereby each PHEV declares a willingness-to-pay (WTP) parameter which, in combination with the published price, allows them to determine their individual charging strategy. The grid capacity, therefore, is allocated in proportion to the WTP parameters of each PHEV.

Alternatively, a non-price-based signal can be provided by the grid-responsible party (DSO or other party) such that the optimization is no longer economically based, but with respect to some other objective such as grid capacity. A concept was depicted by Li *et al.* (2011) where the state of charge (SOC) is compared with a charging reference set by the grid-responsible party and binary charge decisions are then made locally at each EV. The charging reference is based on a number of inputs, such as the state of the power system, the underlying nonflexible household demand, and local generation from distributed sources. The objective of the charging coordination is to reduce the demand variance at the distribution substation level, thereby achieving valley filling at a local level. This objective is valid at a local level as it maximizes the utilization of the local network capacity; however, as penetrations of intermittent generation increase, demand will be required to follow generation to a larger extent than currently. This will necessitate multiple objective, or bi-level, optimizations to both optimize local network utilization and provide global system balance support.

The aggregation of EVs into fleets for collective centralized control is frequently explored in the literature. Harnessing the spatial and temporal flexibility of charging of a fleet of vehicles aids the prevention of congestion. Galus and Andersson (2009) studied the impact of PHEVs and proposed the implementation of a 'PHEV manager' which manages the charging on a given network, or section thereof. The proposed vehicle manager provides the network operator with predicted charging profiles and the operator ensures there are no system violations. In the case of component overload or other violation, the operator relays load restrictions to the managers who must then redispatch charging to respect these limits. In this manner the DSO is responsible for monitoring the network and ensuring system security. Where congestion occurs, a bidding process takes place, ensuring that only those vehicles with a higher priority for charging maintain their connection. A bidding process such as this is an economically efficient solution to this problem, but requires a high level of monitoring and communication from the network operator to multiple parties. This may be an intelligent approach for nonresidential charging where behavior is less predictable. In a residential setting, it is possible that charging will follow a more regular cycle and such intensive communications will not be required.

The study conducted by Lopes *et al.* (2009) uses centralized control to optimize EV charging so as to maximize the tolerable EV penetration, while respecting the physical grid restrictions. A dual-tariff system is found to increase the tolerable penetration from 10% to 14% versus uncontrolled charging, while smart charging was found to increase this to 52%. In this case, a centrally controlled optimization was simulated where the charging behavior was dictated by the voltage limits at each bus and the apparent power flow limitation on each branch.

Sundstrom and Binding (2010) introduced the concept of a commercial FO. In this case, the FO is responsible for predicting driving requirements and ensuring vehicles are charged with respect to their individual user requirements, while also minimizing the cost of charging. Grid constraints are included directly into the optimization. Linear programming is used to determine the charging schedule for a given 24 h period, with an iterative process employed to alleviate any congestion detected on the network. Load flow analysis is performed on each iteration, and additional constraints are included in the following iteration to deal with congestion issues. Even considering a relatively small network of 50 000 EVs, 200 000 edges, and 100 000 nodes, the number of variables is 25×10^6 and the number constraints is 15×10^6. This case corresponds to a small city; the problem would be far larger for a large city or region; correspondingly, the computational power required to solve this problem would be substantial. There are a number of issues concerning this approach; primary among these is the level of detailed information on the grid required by the FO. In addition, this approach does not facilitate the operation of multiple FOs on the same network, as they would need to have knowledge of each other's optimization, which is neither practical nor commercially viable.

Lopes *et al.* (2011) proposed an alternative which also uses FOs to coordinate charging, while also allowing for the possibility of multiple FOs operating on the same network. Each FO optimizes their fleet charging schedule to maximize utility to the consumer. This schedule is relayed to the DSO who will analyze the total demand for a given network and issue schedule alteration requests to an FO when overload or excessive losses are detected. This is commercially attractive as it allows competition between FOs, in keeping with the principles of a deregulated electricity market. The transfer of information is also minimized, as the FO does not require any knowledge of the grid. A drawback of this approach is that it is not economically efficient: the FOs will require compensation from the DSO for the requested schedule alterations. A pure market mechanism would be a more efficient solution.

Coordinated charging has emerged as a highly effective manner in which to prevent grid congestion. Harnessing the flexible nature of EVs is clearly an effective approach; however, there is no clear agreement on the optimal method by which to achieve effective congestion prevention (Papavasiliou *et al.*, 2011). Real-time optimizations have been explored in the literature (Fan, 2011; Li *et al.*, 2011); however, owing to the dynamic nature of EV driving requirements, the flexibility resource available in real time may be less than if charging is planned from a day-ahead perspective. Day-ahead planning for congestion prevention establishes an optimal charging profile that will result in congestion-free operation if all uncertainties follow the predicted behavior. Deviations from the predicted behavior will require intervention in real time; however, the required correction will likely be less than if charging behavior was not previously established on a day-ahead basis according to the physical constraints of the system.

10.2   Dynamic Tariff Concept

The flexible nature of EV-charging demands lends to their use for the provision of system services; however, any meaningful contribution from EVs for these services requires a significant number of EVs. Such levels of additional demand may not be tolerated on older networks that were designed for conventional nonflexible demand. Without adequate consideration of these grid constraints and preventative action, extensive grid reinforcement and upgrade will be required, at significant expense.

Consequently, the flexibility of EVs should primarily be used for the prevention of congestion through the dispersion of charging away from periods and locations of excessive grid loading. A price signal, or dynamic congestion tariff, is an effective means by which to incentivize such behavior. As congestion is a locational phenomenon, such a tariff would also vary temporally and geographically to reflect the level of system loading at each location and time period. The application of a dynamic tariff (DT) on a day-ahead basis is effective as it allows EVs or aggregators to establish a congestion-free baseline, based on predictions of charging behavior and day-ahead electricity prices. Should any variable deviate from the prediction, congestion may result and additional control will be required closer to the time of operation; however, effective application of tariffs should minimize the need for such intervention.

The framework of such a day-ahead DT can take many forms; however, these can be separated into two distinct categories: integrated and stepwise. An integrated framework involves the implementation of a fully nodal pricing system where EV aggregators or charging service providers submit demand bids to the spot market and are then subject to locational prices. In this case both system balance and grid congestion are settled in a single step. This is considered to be the most socio-economically optimal solution as the locational price of electricity reflects the marginal cost of power supply at each system bus. This nodal pricing or locational marginal pricing approach is used in a number of electricity market worldwide, such as PJM, ISO-NE, and ISO-NY. A form of this is already implemented in the Nordic market through zonal pricing, where transmission system congestion and system balance are dealt with through the process of market splitting. Implementing this on the distribution network, however, would require a complete overhaul of the current day-ahead market and would be difficult to introduce, particularly from the perspective of market participant acceptance.

A stepwise system is one in which the system balance and grid congestion are handled separately. Generally, this requires prediction from one or more market participants: DSOs can predict demand and determine tariffs based on their predictions, or aggregators can predict tariffs and optimize their demand so as to avoid congested periods and locations. There is no fixed method for determining the magnitude of the tariff applied in this concept. It is considered complex to achieve socio-economic optimality in this concept; however, it is very simple to introduce into the spot market as it currently exists.

The DT scheme developed here is based on the concept of a stepwise process where grid congestion is managed prior to the determination of system balance on the day-ahead market. The proposed scheme is intended to fit directly into the Nordic day-ahead electricity market, Nord Pool Spot. This will ensure ease of acceptance from all market participants.

DTs are applied to flexible demand with the objective of incentivizing congestion-preventing behavior; thus, their application to nonflexible demands is inappropriate. The exposure of nonflexible demand to such fluctuating tariffs would unfairly penalize non-EV customers based on the congestion-inducing demand of their neighbors. As current distribution networks were designed for conventional demand, the additional congestion-inducing flexible demand from EVs can be viewed as the marginal demand and should be charged appropriately.

The DT is designed for implementation in addition to the conventional flat tariff for use of the network. If DTs are only applied in cases of congestion, then there is the possibility that the DSO will not generate adequate revenue and the grid will not be sufficiently maintained or reinforced. It is assumed that congestion is not the typical situation (DTs only apply in the case of congestion) and that the combined tariffs are sufficient to drive continued grid reinforcement and maintenance. Regulation of the DSO will be required to ensure that the revenue accrued is appropriately allocated for grid reinforcement and maintenance.

A key concern with a stepwise tariff process is the difficulty in attaining a socio-economically optimal solution. In this scheme, it is proposed that the tariffs will be based on the locational marginal price (LMP) of electricity; thus, the tariffs will reflect the true cost of generation, transmission, and distribution of electricity. The use of the LMP as a basis for the tariff ensures that EV demands at each bus will be penalized in proportion to their individual impact on network congestion.

A nodal pricing market is simulated by the DSO; however, the price is implemented in the format of tariffs for flexible demand (in this case EVs), thus avoiding alterations to the structure of the current spot market.

There will be a compromise between socio-economic optimality at high resolution and demand forecast robustness at lower resolution. At higher resolution the tariffs will vary at each node and will therefore reflect the marginal cost for each node. However, at such a high resolution the reduced reliability of demand forecasting is unlikely to be sufficient to successfully prevent congestion at all times. At lower resolutions, where each tariff will apply to a number of nodes, the forecast reliability will probably be higher, resulting in more accurate tariffs for the prevention of congestion. However, the tariffs will not necessarily reflect the true marginal cost for all customers within the particular price zone.

The concept of DTs is dependent on the predictable and flexible nature of EV loads when charged in a residential setting (or other long-term charging locations). This does not apply to vehicles charged at on-street charging points or other public charging locations. These nonresidential charging stations will charge vehicles at random intervals and possibly for shorter periods than residential charging. Charging at such locations is likely to be dominated by user-defined charging behavior, for instance where parking duration and charging requirements are declared by the vehicle owner upon parking. Prediction of charging behavior at such locations is more complex than that in a residential area and, as such, the application of day-ahead tariffs may be inappropriate here. It is envisaged that an alternative tariff model will be required for such charging points.

Direct response of EV owners to fluctuating tariffs is unlikely to generate the desired response; conventional electricity consumers are subject to electricity payments of which the spot price typically only represents about one-third, so their incentive to behave in a flexible manner is minimal. Instead, it is proposed that charging is overseen by an aggregator, or FO. It is envisaged that the FOs will operate as service providers to consumers who will purchase a particular energy service; for example, battery SOC or traveling distance. In this manner, consumers are insulated from price fluctuations while still ensuring demand flexibility.

As commercial entities, FOs are profit oriented; exposure to fluctuating prices and fixed revenue from EV demands will drive FOs to fully exploit the flexible nature of their demand portfolio to ensure minimum-cost operations. As such, DTs will ensure the most optimal charging behavior for the prevention of congestion.

10.2.1 DT Framework

The DT is applied on a day-ahead basis and is calculated prior to day-ahead market operations. The tariff is issued to FOs in a timely manner such that it can be incorporated into their optimizations and considered during the formulation of bids for the day-ahead market. There are two major stakeholder groups in the DT framework: the DSO and the FOs. The proposed framework allows for multiple FOs on any given network, allowing for competition between charging service providers, in keeping with the convention of a deregulated energy market.

Calculation of a demand-dependent tariff before demand is declared requires the prediction of the locational charging behavior, conventional nonflexible demand, and the day-ahead price profile. These predictions are required by both the DSO and the participating FOs.

Driving requirements of commuter and home-use vehicles will likely follow a regular pattern, varying by weekday or weekend, holiday periods, and possibly by season. Historical driving data can therefore be used to generate the required predictions. A method for the prediction of daily driving requirements was proposed by Aabrandt *et al.* (2012). Conventional demand similarly follows a regular pattern.

The prediction of day-ahead electricity prices will require inputs from a number of sources, including the operating requirements of must-run units such as combined heat and power and base-load plants, and the predicted wind power production level. As penetrations of low marginal cost renewable generators increases, the weather forecast will have an increasing impact on the day-ahead price profile.

The locational charging demand in response to the day-ahead price profile can be calculated once predictions have been determined; assuming that all FOs optimize their charging schedules to achieve the most cost effective charging behavior. The resulting profile is considered the undisturbed charging response; from this charging profile, the incidence of congestion can be determined and the appropriate tariff can be calculated. The flowchart of this process is shown in Figure 10.1.

10.2.2 DT Calculation

Locational marginal pricing ensures economic optimality of the electricity market as it reflects the true cost of the supply of electrical energy at each location in the network; that is, the marginal cost of generation and the costs associated with the physical constraints of the network.

Ideally, any tariff applied to EV charging demands should also reflect the cost of the provision of charging energy. For this reason, the DT is based on the LMP; however, certain adjustments are required to reflect the nature of EV demands, as explained further below.

Electricity markets operating under a nodal pricing structure require the submission of generation and demand curves from market players to the day-ahead market. These curves are used as inputs to an optimal power flow (OPF) algorithm; this algorithm goes beyond the traditional generation-demand balance constraints to incorporate those constraints associated with line flow, bus voltage, and reactive power flow, among others.

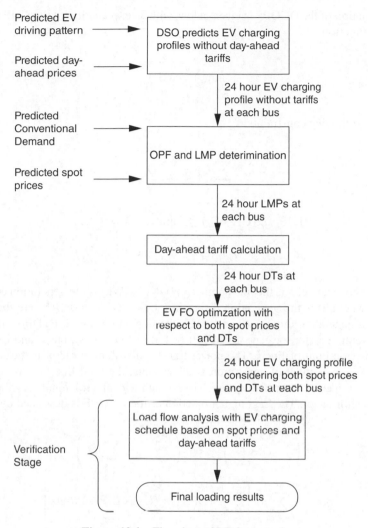

Figure 10.1   Flowchart of DT framework.

The objective function selected here is a generation cost minimization; and as demand is considered inelastic, this corresponds to social welfare maximization. Conventional OPF calculations may use other objectives, such as a minimization of power loss.

DC power flow is a useful tool for the rapid calculation of real power flows; however, no information is provided on reactive power flow and voltage is assumed to be unity per unit at each bus. As congestion commonly manifests in the form of line and transformer overload, the DCOPF algorithm is considered appropriate for the calculation of LMPs in this case. The increased speed of calculation is a major advantage for this application owing to the large number of nodes present in the distribution network. As the tariff for EV demand should find its basis in the cost of the congestion that it induces, the cost of losses is not considered relevant; as such, a lossless DCOPF is used.

The formulation of the DCOPF is given below, with an explanation of the LMP calculation. Objective function

$$\min \sum_{i=1}^{N} c_i(G_i) \tag{10.1}$$

subject to the following constraints:

$$\sum_{i=1}^{N} G_i = \sum_{i=1}^{N} D_i \tag{10.2}$$

$$\sum_{i=1}^{N} \text{PTDF}_{k-i} x(G_i - D_i) \leq \text{Limit}_k, \quad k = 1, 2, \ldots, M \tag{10.3}$$

$$G_i^{\min} \leq G_i \leq G_i^{\max} \tag{10.4}$$

where n is the bus number, i is the line number, c_i (DKK/MWh) is the generation cost function at bus i, G_i (MWh/h) is the generation dispatch at bus i, $G_i^{\max}$ and $G_i^{\max}$ are the maximum and minimum generation output at bus i, D_i is the demand at bus i, PTDF_{k-i} is the power transfer distribution factor, or generation shift factor, to line k from bus i, and Limit_k (MW) is the transmission limit of line k. The power transfer distribution factor is the sensitivity of a line flow to an increment of generation at a bus; thus, the total line flow is comprised of contributions from generation at all system busses, each with an individual contribution factor, or generation shift factor. The Lagrangian function of the DCOPF is described by

$$L = \left(\sum_{i=1}^{N} c_i \times G_i \right) - \lambda \times \left(\sum_{i=1}^{N} G_i - \sum_{i=1}^{N} D_i \right)$$
$$- \sum_{k=1}^{M} \mu_k \left[\sum_{i=1}^{N} \text{PTDF}_{k-i} \times (G_i - D_i) - \text{Limit}_k \right] \tag{10.5}$$

The LMP at bus i can be calculated using

$$\text{LMP}_i = \frac{\partial L}{\partial D_i} = \lambda + \sum_{k=1}^{M} \mu_k \times \text{PTDF}_{k-i} \tag{10.6}$$

The LMP formulation shown above results in nodal prices that are related to the bidding curves of the system generators. Applying a DT based on this formulation would result in an ineffective EV charging response as the signal would be more related to generator bidding preferences than EV charging behavior. The resulting LMP would suitably result in tariffs that cover the duration of congestion; however, the magnitude would not be appropriate to generate the required response. As an example, consider the case where the price range over 24 h is between 200 and 500 DKK/MWh and congestion is present at the lowest cost hour;

a tariff of 20 DKK/MWh is unlikely to produce any meaningful congestion prevention as the price at the trough will remain among the lower cost hours. Thus, the LMP formulation for the purposes of tariff calculation must take into account the prices at all time periods, rather than focus on the conditions of a single time period.

For this reason, an alternative formulation is used for the purpose of generating tariff values. Under this alternate formulation, the system is modeled with only two generators: the primary generator supplies power at the predicted market clearing price and the secondary generator supplies power at the market clearing price plus a premium, where the premium is equal to the maximum range of prices over the optimization period. The generators are placed on opposite ends of the network, to facilitate power flow redirection and congestion alleviation by the secondary generator.

Under congestion-free conditions, the primary generator will supply all the required power at the market clearing price, and the tariff will be zero. During congested periods, the secondary generator will be required to generate power so that the line flow constraints of the DCOPF are respected. The tariff resulting from this formulation will reflect the daily price profile, and thus its magnitude will be sufficient to generate the required charging response, while still indicating the relative contribution of each EV to system congestion.

In addition to the above-mentioned alterations, consideration must also be given to the line flow limit $Limit_k$. If this limit is kept as the actual physical limit, the resulting tariffs will effectively shift charging away from that congested period; however, if the adjacent time periods are also low-cost hours, the charging will simply be shifted there. These periods are likely to have been highly loaded before the shifting, and the additional charging will result in congestion where none was previously (i.e., a secondary congestion peak). Thus, it is important that the time extent of the LMPs ensures that charging is adequately shifted among periods of low system loading. This shifting can be partially accounted for by reducing the line flow limit employed in the DCOPF calculation to further disperse charging away from period of excessive system loading.

In some cases it is possible to identify where secondary congestion may emerge following the application of tariffs; however, this is an uncertain approach and may result in ineffective or unnecessary tariffs. Evidently, an iterative approach is necessary to ensure the optimum application of tariffs.

The approach adopted here is to set the line flow limit used in the DCOPF calculation so that the resulting line flow limit represents the mean system loading resulting from conventional demand alone. If the resulting tariff were applied directly it would result in excessive tariffs, as it would cover almost the entire optimization period. Instead, the system loading curve is considered: a curve of the maximum loading on any component on the system. The system loading curve is considered in conjunction with the LMP, and tariffs are only applied where system loading is above a given level (discussed further below).

This initial application of tariffs is likely to result in secondary congestion peaks; thus, a second iteration of the tariff application process is required. As the DSO is aware or has predictions of the charging requirements and constraints and can determine the charging response to a given price profile, this does not require any iterations involving the market. All iterations can be run by the DSO alone, and the final tariff can then be issued to the FOs. Another consideration required here is that, on further iterations, congestion may be shifted back to locations where tariffs were applied on previous iterations; this indicates that a higher tariff magnitude is required.

We can consider here that the tariff is comprised of two elements: the magnitude and time extent. The magnitude is determined by the altered LMP, whereas the time extent is handled through an iterative analysis of the system loading curve. This process is illustrated in the flow chart in Figure 10.2.

There are two variable quantities included in this optimization: the system loading level above which tariffs are applied X and the tariff increase for repeated congestion Y. This selected loading level X can be varied to improve the accuracy of tariff application: a higher level (up to 100%) will result in more accurate tariff application, but more iterations are required to fully alleviate congestion; a lower level will result in a reduced number of iterations, but the tariff application may not be optimal. The same reasoning applies to the tariff increase, where smaller tariff increments result in a more accurate solution, but increased iterations are required. The values of X and Y can be tuned to meet user preferences.

10.2.3 Optimal EV Charging Management

The concept of smart charging, or strategically scheduled charging, is one that is frequently discussed in the literature in this area, as detailed in Section 10.1. It is widely agreed that FOs or aggregators will schedule charging so as to minimize their total cost. This cost-minimizing optimization is in line with their profit-seeking objectives as a commercial entity. The DT concept outlined in Section 10.2.2 is dependent on the assumption that FOs will optimize the charging behavior of their fleet in response to a price signal, in the form of the combined day-ahead price and dynamic grid tariffs.

An optimal EV charging profile can be found through the use of a linear programming optimization, as shown below. The use of linear programming in this case assumes that the FO is a price taker and does not exert market power. This is a valid assumption in the early stages of EV adoption; however, as fleets increase in size it is possible that a measure of market power may be wielded by FOs. Kristoffersen *et al.* (2011) detail how this could be incorporated into a similar optimization through the use of quadratic programming; a quadratic objective function can be used to allow for changes in the day-ahead market due to gaming from FOs.

The mathematical formulation of the optimal charging optimization for FOs is shown below; the objective function ensures that the total cost of charging is minimized with respect to the combination of the predicted day-ahead price and the published locational DT.

Objective function

$$\min \sum_{t=1}^{T} \sum_{i=1}^{I} \sum_{n \in i} (E_{n,t}(C_t + \mathrm{DT}_{i,t})) \tag{10.7}$$

subject to the following EV charging constraints

$$0 \leq E_{n,t} u_{n,t} \leq E_{\max} \quad \forall t, n \tag{10.8}$$

$$\mathrm{SOC}_{\min} \leq \mathrm{SOC}_{\text{init}} + \sum_{t=1}^{\tau_n} E_{n,t} u_{n,t} - \sum_{t=1}^{\tau_n + \tau_{n,d}} E_{d,n,t} v_{n,t} \leq \mathrm{SOC}_{\max} \tag{10.9}$$

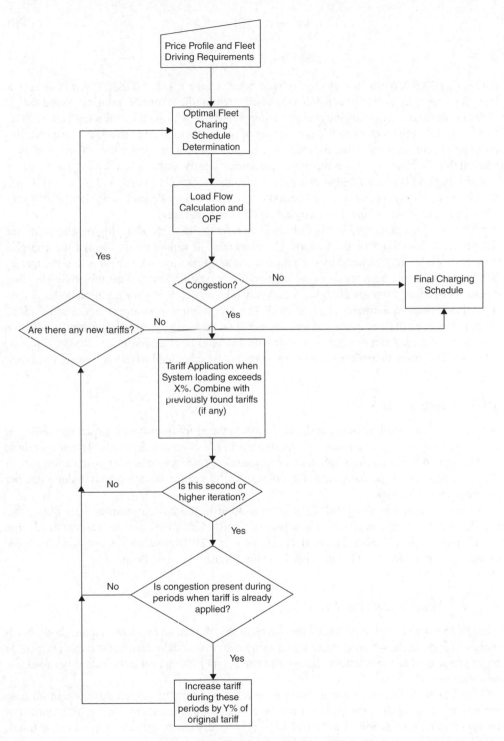

Figure 10.2   Iterative tariff application.

$$u_{n,t} + v_{n,t} \le 1 \quad \forall t, n \tag{10.10}$$

$$v_{n,t}, u_{n,t} \in \{0, 1\} \quad \forall t, n \tag{10.11}$$

where C_t (DKK/kWh) is the electricity spot price at time t, $DT_{i,t}$ (DKK/kWh) is the DT at bus i for time t, $E_{d,n,t}$ (kWh) is the driving energy requirement for EV n during period t, E_{max} (kWh) is the maximum charging energy during period t, SOC_{min} is the minimum battery SOC, SOC_{init} is the initial battery SOC, ι_n is the set of times at which a vehicle leaves the charging point and becomes unavailable for charging, $\tau_{n,d}$ is the set of durations for which a vehicle is unavailable for charging, $v_{n,t}$ is the binary parameter signifying the EV's driving status, where $1 = $ driving and $0 = $ available to charge, i is the bus index, n is the EV index, $E_{n,t}$ (kWh) is the charging energy for EV n during period t, $u_{n,t}$ is the binary control variable signifying the EV's charging status, where $1 = $ charging and $0 = $ not charging.

The constraints employed in this optimization refer to the users' charging requirements and the physical limitations of the battery. The constraint in Equation 10.8 limits the charging power; this is usually dictated by the grid connection. Equation 10.9 ensures that the battery SOC upon departure from the charging location is at least sufficient to meet the driving requirements while away from the charging location, while also allowing for additional charging to meet the driving requirements at a later time. This constraint also ensures that the battery SOC remains between the minimum and maximum limits; these are required to prevent a reduction in battery lifetime from excessive charging or discharging (Yoshida *et al.*, 2003). Equation 10.10 is a constraint that ensures the charging cannot be scheduled when the vehicle is driving.

10.3 Case Studies

A number of case studies were conducted to assess the efficacy of the proposed dynamic grid tariff concept for the prevention of congestion from EV charging demands. The aim of these studies was to determine the limitations of the concept, the key congestion-inducing factors are identified and the performance of the dynamic grid tariff is demonstrated under a number of congestion scenarios.

As the concept was designed for implementation in the Nordic market, it is fitting that the study environment in use here is based on Danish data. Driving data were sourced from the Danish National Travel Survey (DNTS; Wu *et al.*, 2010), and the network on which EV charging is simulated is an LV network from the Danish island of Bornholm.

10.3.1 Vehicle Driving Data

The DNTS is a rich source of data on the driving habits of Danish residents. Danish households were surveyed on their driving behavior on a single particular day; thus, the survey data provide an impression of driving patterns across all vehicle types, not a particular category of vehicle or consumer.

The data from this survey were used to develop a cohort of vehicles for use in all case studies. A subset of the provided data was identified as most pertinent for the development of an accurate representation of a fleet of EVs: driving distances, driving stop and start times, stopover locations and durations, and day of the week.

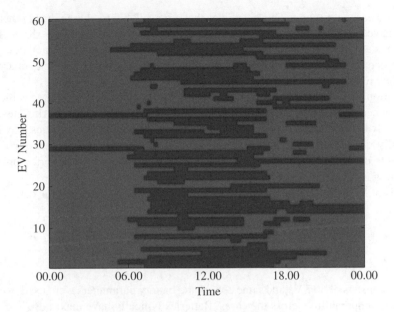

Figure 10.3   Fleet availability to charge.

As the DNTS was conducted on conventional vehicles, certain samples are not suitable for analysis as EVs as they travel distances beyond the capabilities of a typical EV. Such samples were filtered from the data, with the consideration that all driving energy requirements should be met while respecting the limitations placed by battery capacity and possible charging durations. Further details on battery capacities are given below.

Driving behavior naturally varies with time, particularly between working and holiday days; in order to ensure an accurate representation of driving behavior, the data were filtered to select only those samples taken on a particular day of the week. For the case studies that follow, all driving data are from a Monday. A cohort of 60 EVs was selected at random from the filtered data.

The travel distance of the finalized EV cohort was converted to an energy requirement through the use of a conversion factor; as in Wu *et al.* (2010), a factor of 0.15 kWh/km was used. The finalized vehicle cohort was primarily used for demonstration of the DT concept; however, analysis of the driving energy requirements and availability to charge of the cohort also provided insight into the incidence of congestion. This analysis was valuable for the development of case studies, particularly the generation of price profiles to test the performance of the DT concept. Three factors were identified as highly influential on the incidence of congestion: level of availability to charge within the cohort, charging requirements during the availability window, and the price profile. Figure 10.3 shows the availability of the selected cohort for charging during a given 24 h period. Each EV is represented by a horizontal section, and the availability to charge is indicated in white, driving periods are indicated in grey. In this case, vehicles are only available to charge when at the primary residence.

It can be seen from Figure 10.3 that the study cohort is comprised of both commuter vehicles and short-journey home-based vehicles. This results in low charging volumes during the late morning and early afternoon. Those vehicles that are available to charge during these periods typically have low charging requirements as their driving distances are short. Therefore, congestion is unlikely during these periods, even when the price profile is such that the lowest cost hours occur in the middle of the day, when many vehicles are available to charge, simply because they may not need to charge.

Observation of Figure 10.3 informs the reader that many of the commuter vehicles return to the residence in the early evening, around 17:00, as is typical in Denmark. These vehicles have typically been driven further than home-based vehicles and their batteries may have been substantially depleted, resulting in a large charging requirement to prepare for the driving requirements of the following morning. For this reason, a trough in the price profile in the early evening induces a substantial level of congestion on the test network.

Analysis of the driving data informed the selection of price profiles for the case studies that follow.

10.3.2   EV Cohort Characteristics

The cohort consists of 60 EVs; the same cohort and battery parameters were used in each study to maintain comparability across the cases. Batteries typically have maximum and minimum SOC limitations; here, they are assumed to be 85% and 20% of battery capacity respectively. The maximum charging power of each EV is assumed to be 2.3 kW; this is not determined by the battery model, but rather the grid connection.

The battery capacity and initial SOC varies by EV according to the driving energy requirements; the mean battery capacity is 20 kWh and the mean initial SOC is 50%. The final SOC is determined by the driving requirements and charging behavior during the day, this corresponds to the initial SOC for the following day.

As the charging schedule optimization is a cost minimization and there is no limitation on the final SOC, the cohort will tend towards battery depletion. A number of approaches can be used to counteract this, such as specifying a minimum final SOC. This approach does not consider the possibility of more economical charging at a later time to meet the driving energy requirements of the following day. In the case studies that follow, the approach is taken to run the optimization over 48 h, but to consider only the first 24 h for submission to the day-ahead market. This prevents widespread battery depletion, as charging can be scheduled with respect to prices and driving requirements in periods that were previously outside of the optimization window.

10.3.3   Price Profiles

Each case study is conducted on an identical cohort of EVs; however, a unique price signal is supplied in each case. The price profiles were selected based on results from the analysis of driving data described previously. They were selected with the objective of illustrating the effectiveness of the DT concept under a variety of price profile stimuli. A control case with a constant price over the entire optimization period is used to illustrate that congestion is not the normal operating condition, but rather that it is induced by price stimuli that cause a concentration of charging resulting in network congestion.

Table 10.1   Case study price profile details.

Case study	Price profile details
1	Flat profile (control)
2	Double trough – early morning and evening
3	Single peak – night
4	Single peak – evening
5	Single trough – night
6	Single trough – evening (worst-case scenario)
7	Fluctuations throughout optimization window
8	High-frequency fluctuations throughout optimization window

The remaining cases use strategically generated price profiles. As the previous discussion on driving data highlighted, congestion is most likely to emerge during low-cost hours with high vehicle availability following extended periods of travel. A worst-case scenario is developed with the lowest cost hours placed in the early evening, as commuters return home. Additional single-peak or single-trough price profiles demonstrate the congestion that would be induced without the application of the DT. Multi-peak and -trough price profiles are also employed in the case studies to investigate the charging response to more volatile price signals.

The details of the price profiles are shown in Table 10.1 and in the figures that follow.

10.3.4   Electrical Network

A 0.4 kV distribution network from the Danish island of Bornholm was selected for the case studies. The network is shown in Figure 10.4. It consists of two radial sections connected by a single branch. The network has 33 busses, 33 cables, 72 residential customers, two 10/0.4 kV transformers and two generators. The generator located at the top of Figure 10.4 represents power flow from the external grid; the second generator is a simulation tool used to determine the LMP. This second generator facilitates reverse power flow along congested branches in the OPF formulation. The generators are located on opposite ends of the network to facilitate maximum power redirection. The DTs are determined from the LMP calculation; once these are correctly applied for congestion alleviation the second generator does not supply any power, as there is no requirement for power redirection.

10.3.5   Software and Case Study Parameters

The case studies were conducted using Mathworks' MATLAB® and DigSilent's PowerFactory. The linear programming package in MATLAB® was used to calculate the optimal charging schedules. PowerFactory was used to model the network and to calculate the LMPs using the DCOPF solver.

The DT formulation, as described in Section 10.2.2, requires the tuning of the X and Y parameters. In the case studies here the X value was taken as 90% and the Y value as 25%. These values provided satisfactory results within a few iterations in each case.

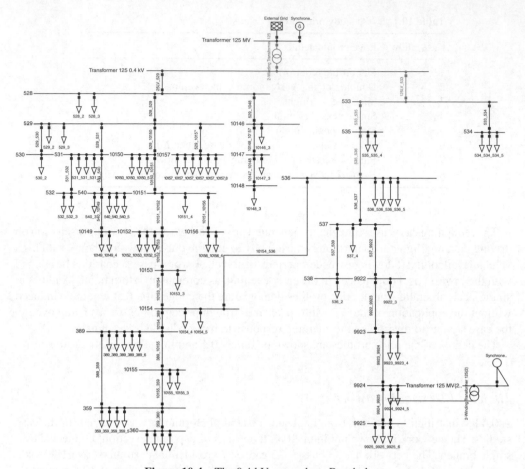

Figure 10.4   The 0.4 kV network on Bornholm.

10.3.6   Case Study Results

The figures shown below illustrate the system loading with and without the application of DTs. System loading is used as an indicator for the level of congestion present on the network. As component loading varies throughout the network, the system loading indicator represents the maximum real power loading on any line or transformer on the system; system loading in excess of 100% indicates congestion. The day-ahead price profile is shown to inform the reader of the impact of the price profile on the incidence of congestion. The dynamic grid tariffs are also shown; the mean value of the locational tariff is shown for ease of comprehension.

The control case is presented in Figure 10.5 to illustrate that congestion is not present in the absence of a varying price stimulus. It can also be concluded from this case that it should be possible to fully alleviate congestion through the application of appropriate price stimuli to disperse charging.

It should be noted that this control case does not represent uncontrolled charging, or "dumb-charging," as EVs are only charged to meet their driving requirements, whereas under an

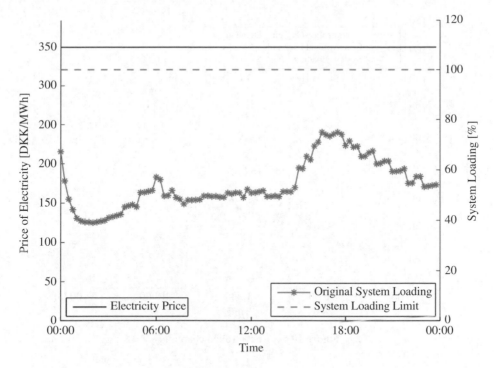

Figure 10.5   Case 1 – control.

uncontrolled charging scenario vehicle batteries would be charged fully. A representation of uncontrolled charging of the case study vehicle cohort is shown in Figure 10.6; for this analysis it is assumed that all vehicles are fully charged prior to the start of the optimization period. This is a conservative assumption; alternative selections of the initial battery SOC would have resulted in additional charging in the early morning period but would not impact on the congestion peak as this is caused by charging following battery depletion from day-time driving.

A time series of the system loading from conventional demand is included in this figure to further illustrate that minimal charging occurs during the day, and the majority of charging demand appears to come from commuter vehicles.

Case 2, shown in Figure 10.7, illustrates one of the more challenging scenarios, where low-cost hours occur during both the evening and morning, prime periods for commuter vehicle charging. Excessive system loading of 147% is successfully reduced to 100%. Should it be necessary to reduce the system loading further, the tariff calculation algorithm can be provided with an alternative congestion threshold.

High tariff levels are required to fully alleviate congestion, however tuning of the X and Y parameters in the DT algorithm may reduce this cost as tariffs can be applied at a finer resolution.

Case 3, shown in Figure 10.8, demonstrates a case where congestion is more easily alleviated through application of tariffs over the peak periods only.

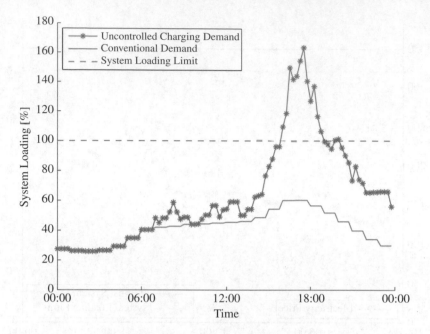

Figure 10.6   Uncontrolled charging.

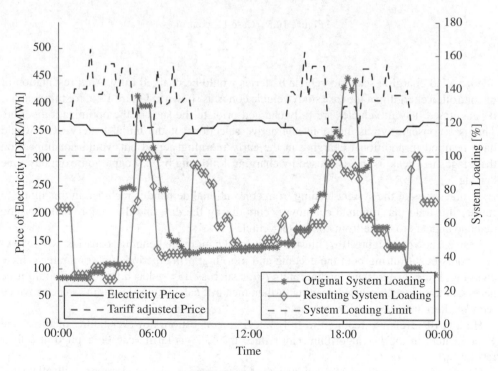

Figure 10.7   Case 2.

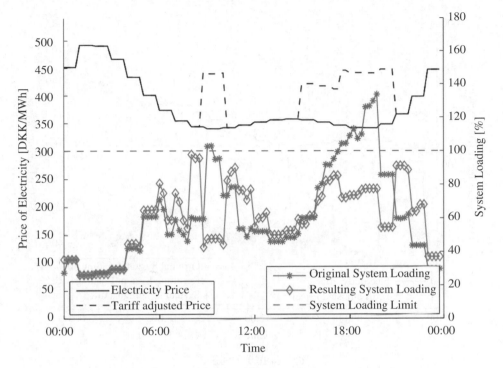

Figure 10.8 Case 3.

Case 4 (Figure 10.9) illustrates one of the simplest cases of congestion alleviation. Here, the price peak coincides with the typical evening charging peak, and charging is consequently scheduled outside of this period. Only slight congestion is evident, and this can be successfully alleviated through the application of DTs. The resulting system loading during these periods is very low, and the resulting system loading results from conventional demand only as EV charging is completely shifted from these periods.

Case 5 (Figure 10.10) illustrates the impact of a price trough during the night. This period is considered a prime charging period. Currently, many electricity providers offer dual-tariff pricing, with reduced prices during the night, this case illustrates that this may not be suitable for EVs. Initially, it appears that congestion only occurs in the early morning period, but the application of DTs in the late evening indicates that congestion is shifted to these periods during subsequent iterations of the DT algorithm.

Case 6 (Figure 10.11) represents the worst-case scenario, where a price trough in the early evening results in congestion approaching 160%. As mentioned before, this level of congestion is induced by the coincident peaks of charging requirements and EV availability, and the trough in day-ahead price profile. Nonetheless, this congestion is easily alleviated through the application of DTs.

Cases 7 and 8 (Figure 10.12 and Figure 10.13 respectively) represent some of the easiest cases for congestion alleviation. The frequent fluctuations in price result in the concentration of charging into short periods, initially causing congestion; however, only slight alterations

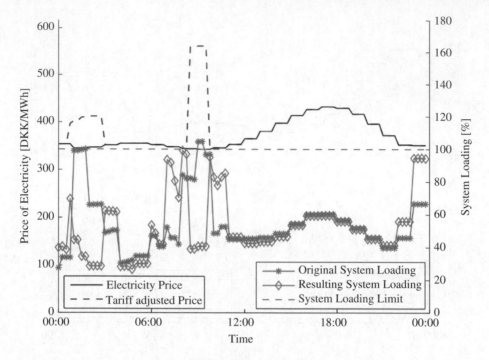

Figure 10.9   Case 4.

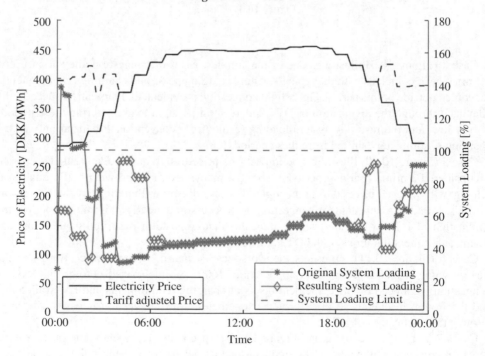

Figure 10.10   Case 5.

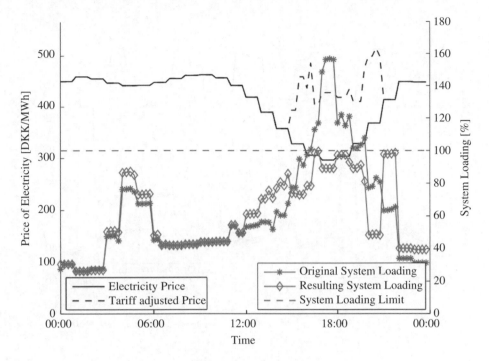

Figure 10.11   Case 6.

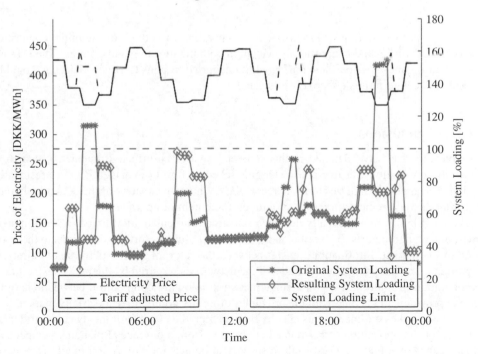

Figure 10.12   Case 7.

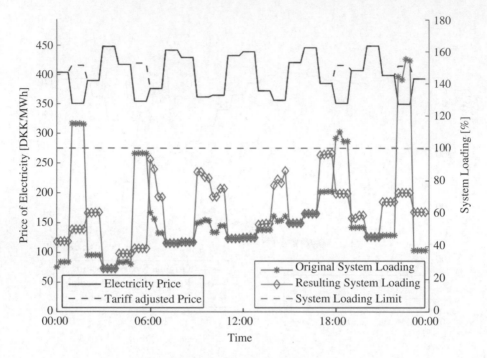

Figure 10.13 Case 8.

of the price profile are required to disperse this charging. In both cases the initial congestion levels exceed 150%, and congestion is evident in both the morning and evening. The DTs are applied during these periods, and during additional price troughs to which charging would be shifted, and congestion is effectively alleviated.

10.4 Conclusions

The dynamic grid tariff detailed in this chapter provides an effective manner by which to establish a congestion-free day-ahead charging schedule for FO-managed EVs in a residential setting. The application of this tariff increases the EV hosting capacity of the network, thereby deferring the requirement for costly system reinforcement or upgrade.

The tariff has been specifically designed for direct implementation in a day-ahead market environment, particularly the Nordic market, without requiring further alterations to the operation of the market. Information transfer between the relevant market players is minimized; in particular, no knowledge of the status of the network is required by the FO. This facilitates the operation of multiple FOs on a single network, allowing competition between charging service providers in keeping with the principles of a fully competitive electricity market.

The case studies outlined in this chapter demonstrate the efficacy of the dynamic grid tariff concept for the prevention of EV-related congestion from a day-ahead planning perspective. It has been shown that the DTs result in successful congestion prevention in all of the cases examined; however, it should be noted that in certain cases very high tariffs are required to ensure full congestion prevention. In such cases an assessment of the social cost of the tariffs and the benefit arising from congestion prevention should be conducted. The prevention of excessive DTs may form a component of the regulatory rules imposed on the DSO as a natural monopoly.

References

Aabrandt, A., Andersen, P.B., Pedersen, A.B. *et al.* (2012) Prediction and optimization methods for electric vehicle charging schedules in the EDISON project, in *2012 IEEE PES Innovative Smart Grid Technologies (ISGT)*, IEEE, Piscataway, NJ.

Dyke, K.J., Schofield, N., and Barnes, M. (2010). The impact of transport electrification on electrical networks. *IEEE Transactions on Industrial Electronics*, **57** (12), 3917–3926, doi: 10.1109/TIE.2010.2040563.

Fan, Z. (2011) Distributed charging of PHEVs in a smart grid, in *2011 IEEE International Conference on Smart Grid Communications (SmartGridComm)*, IEEE, Piscataway, NJ, pp. 255–260.

Galus, M. and Andersson, G. (2009) Integration of plug-in hybrid electric vehicles into energy networks, in *2009 IEEE Bucharest PowerTech*, IEEE, Piscataway, NJ, pp. 1–8, doi: 10.1109/PTC.2009.5282135.

Gong, Q., Member, S., Midlam-Mohler, S. *et al.* (2011) Study of PEV charging on residential distribution transformer life. *IEEE Transactions on Smart Grid*, **3** (1), 404–412.

Hammerstrom, D.J., Investigator, P., Ambrosio, R. *et al.* (2007) Pacific Northwest GridWise TM Testbed Demonstration Projects Part I. Olympic Peninsula Project.

Hamoud, G. and Bradley, I. (2004) Assessment of transmission congestion cost and locational marginal pricing in a competitive electricity market. *IEEE Transactions on Power Systems*, **19** (2), 769–775.

Kelly, F.P., Maulloo, A.K., and Tan, D.K.H. (1998) Rate control for communication networks: shadow prices, proportional fairness and stability. *Journal of the Operational Research Society*, **49** (3), 237–252, doi: 10.1038/sj.jors.2600523.

Kelly, L., Rowe, A., and Wild, P. (2009) Analyzing the impacts of plug-in electric vehicles on distribution networks in British Columbia. *2009 IEEE Electrical Power & Energy Conference (EPEC)*, IEEE, Piscataway, NJ, pp. 1–6, doi: 10.1109/EPEC.2009.5420904.

Kempton, W. and Tomić, J. (2005) Vehicle-to-grid power fundamentals: calculating capacity and net revenue. *Journal of Power Sources*, **144** (1), 268–279, doi: 10.1016/j.jpowsour.2004.12.025.

Kristoffersen, T.K., Capion, K., and Meibom, P. (2011) Optimal charging of electric drive vehicles in a market environment. *Applied Energy*, **88** (5), 1940–1948, doi: 10.1016/j.apenergy.2010.12.015.

Kumar, A., Srivastava, S., and Singh, S. (2004). A zonal congestion management approach using real and reactive power rescheduling. *IEEE Transactions on Power Systems*, **19** (1), 554–562.

Li, Q., Cui, T., Negi, R. *et al.* (2011) On-line decentralized charging of plug-in electric vehicles in power systems, arXiv:1106.5063v1.

Lopes, J.A.P., Soares, F., and Almeida, P.M.R. (2009) Identifying management procedures to deal with connection of electric vehicles in the grid, in *2009 IEEE Bucharest PowerTech*, IEEE, Piscataway, NJ, pp. 1–8, doi: 10.1109/PTC.2009.5282155.

Lopes, J.A.P., Soares, F.J., and Almeida, P.M.R. (2011) Integration of electric vehicles in the electric power system. *Proceedings of the IEEE*, **99** (1), 168–183.

Maitra, A., Kook, K., and Taylor, J. (2010) Grid impacts of plug-in electric vehicles on Hydro Quebec's distribution system, in *2010 IEEE PES Transmission and Distribution Conference and Exposition*, IEEE, Piscataway, NJ, pp. 1–7, doi: 10.1109/TDC.2010.5484352.

Papavasiliou, A., Oren, S.S., and O'Neill, R.P. (2011) Reserve requirements for wind power integration: a scenario-based stochastic programming framework. *IEEE Transactions on Power Systems*, **26** (4), 2197–2206.

Shao, S., Zhang, T., Pipattanasomporn, M., and Rahman, S. (2010) Impact of TOU rates on distribution load shapes in a smart grid with PHEV penetration, in *2010 IEEE PES Transmission and Distribution Conference and Exposition*, IEEE, Piscataway, NJ, pp. 1–6.

Singh, H., Hao, S., and Papalexopoulos, A. (1998) Transmission congestion management in competitive electricity markets. *IEEE Transactions on Power Systems*, **13** (2), 672–680.

Sundstrom, O. and Binding, C. (2010) Planning electric-drive vehicle charging under constrained grid conditions, in International Conference on Power System Technology POWERCON2010, Hangzhou, China.

Sundstrom, O. and Binding, C. (2011) Flexible charging optimization for electric vehicles considering distribution grid constraints. *IEEE Transactions on Smart Grid*, **3** (1), 26–37.

Taylor, J., Maitra, A., and Alexander, M. (2009) Evaluation of the impact of plug-in electric vehicle loading on distribution system operations, in *2009 IEEE Power and Energy Society General Meeting (PESGM 2009)*, IEEE, Piscataway, NJ, pp. 1–6.

Taylor, J., Maitra, A., Alexander, M. *et al.* (2010) Evaluations of plug-in electric vehicle distribution system impacts, in *2010 IEEE Power and Energy Society General Meeting*, IEEE, Piscataway, NJ, pp. 1–6.

Wu, Q., Nielsen, A., Østergaard, J. *et al.* (2010) Driving pattern analysis for electric vehicle (EV) grid integration study, in *2010 IEEE PES Innovative Smart Grid Technologies Conference Europe (ISGT Europe)*, IEEE, Piscataway, NJ, pp. 1–6. doi: 10.1109/ISGTEUROPE.2010.5751581.

Yoshida, H., Imamura, N., and Inoue, T. (2003) Capacity loss mechanism of space lithium-ion cells and its life estimation method. *Denki Kagaku oyobi Kogyo Butsuri Kagaku*, **71** (12), 1018–1024.

Zhao, L., Prousch, S., Hubner, M., and Moser, A. (2010) Simulation methods for assessing electric vehicle impact on distribution grids, in *2010 IEEE PES Transmission and Distribution Conference and Exposition*, IEEE, Piscataway, NJ, pp. 1–7.

11

Impact Study of EV Integration on Distribution Networks

Qiuwei Wu, Arne Hejde Nielsen, Jacob Østergaard and Yi Ding
Centre for Electric Power and Energy (CEE), Department of Electrical Engineering, Technical University of Denmark, Lyngby, Denmark

11.1 Introduction

The deployment of a large number of electric vehicles (EVs) has become a very interesting option. Replacing conventional internal combustion engine vehicles with EVs will reduce greenhouse gas emissions from the transport sector. In the meantime, the flexibility of EV charging demands can be used to balance the intermittency of renewable energy sources (RES). Although the EV grid integration is beneficial to the environment and can help integrate more RES into power systems, the impact of EV grid integration has to be investigated in order to identify the bottlenecks of power systems for the EV grid integration and test different charging scenarios.

Congestion from EVs can be observed at the medium-voltage (MV) level, as a number of studies demonstrate [1, 2]. Many studies have been conducted analyzing congestion issues on the MV network; however, they also note that the problems likely originate on the low-voltage (LV) network. As such, analysis of this network should be conducted as the primary stage of congestion studies [1, 3, 4].

The degree of grid congestion is dependent on a number of factors, including local grid rating and topology, penetration and distribution of EVs, and charging management procedures. Coordinated charging appears to be an effective method of increasing the penetration of EVs without violating grid constraints. There is some incongruity regarding the optimal manner in which to coordinate charging, with a number of different objectives proposed, including maximization of EV penetration [1], minimization of losses [2], and minimization of customer charging costs [5, 6]. The study conducted by Sundstorm and Binding [5] shows that substantial

Grid Integration of Electric Vehicles in Open Electricity Markets, First Edition. Edited by Qiuwei Wu.
© 2013 John Wiley & Sons, Ltd. Published 2013 by John Wiley & Sons, Ltd.

computational power is required to handle grid constraints in an iterative optimization for EV charging management.

Another method of grid congestion prevention is the inclusion of devices in chargers that detect voltage and halt charging when the voltage drops beyond a given threshold. Alternatively, the power factor could be adjusted to rectify the voltage drop. These methods are mentioned in a number of papers [7–9].

In this chapter, the impact study of EV integration has been carried out in order to quantify the effects of EV demands of different EV charging scenarios on electric component loading and voltage drop, and to identify the bottlenecks within distribution networks. The rest of the chapter is arranged as follows: the impact study methodology and scenarios are presented in Section 11.2; the Bornholm power system is described in Section 11.3; the demand profile modeling is presented in Section 11.4, the impact study results of 0.4 kV, 10 kV, and 60 kV grids of the Bornholm power system are presented in Sections 11.5, 11.6, and 11.7, respectively; conclusions are drawn in Section 11.8.

11.2   Impact Study Methodology and Scenarios

The impact study of EV integration comprises daily time-series power-flow studies using a grid model, existing demands, and EV charging demands in order to investigate the effects of EV integration on electric component loading and voltage drops within distribution networks. The steps of the EV grid impact study are illustrated in Figure 11.1.

11.2.1   Grid Model for EV Grid Impact Study

The ideal scenario for the EV grid impact study is to carry out time-series power-flow studies with the full grid model. However, this is quite difficult in reality. Firstly, it is very difficult to obtain all the data needed to develop a full grid model. For EV charging at home or working

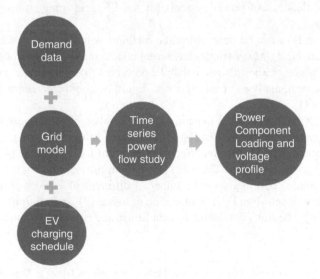

Figure 11.1   Flowchart of EV grid impact study.

places, the EVs will be connected to the LV grid (0.4 kV in Denmark). Therefore, all the LV grids should be modeled. In most cases, distribution system operators (DSOs) do not have all the data for LV grids in a digital format. It will be very time consuming and tedious to convert the available data into a digital format and build the full model of LV grids. Secondly, the LV and MV grids are quite independent from each other during normal operation. Therefore, it is feasible to use typical LV and MV grids to carry out the EV grid impact studies.

Therefore, for the EV grid impact study in this Chapter, the studies have been carried out at the three voltage levels (60 kV, 10 kV and 0.4 kV) with the Bornholm power system models in PowerFactory. For the study at the 60 kV level, the full 60 kV Bornholm grid model with generators and lumped demand at 60/10 kV substations is used to carry out the studies. For the 10 kV and 0.4 kV studies, two typical 10 kV grid models and a typical 0.4 kV grid model have been used to carry out the studies. The selected 10 kV grids are the Ronne Syd 10 kV grid in the downtown area and the Svaneke 10 kV grid in the rural area. The selected 0.4 kV grid is the 0.4 kV grid connected to a 10/0.4 kV secondary substation.

11.2.2 Demand Data

From the power component loading and voltage profile perspectives, the maximum loading and minimum voltage should be obtained and be checked as to whether they are within the specified limits. Therefore, the daily demand with the maximum peak demand has been selected to carry out the time-series power-flow studies.

For the impact study at the 60 kV grid, the measurement data from the Bornholm SCADA system have been used.

For the 10 kV grid, the customer types of all customers within the grid are obtained. The typical demand profiles of different customer types, the customer types obtained, and the yearly electricity consumption are used to get the daily demand profile with 1 h resolution. Afterwards, the demands are summed and put at the 10/0.4 kV secondary substation.

For the 0.4 kV grid, the same method used for 10 kV grid is used to get the daily demand profile with 1 h resolution.

11.2.3 EV Demand Data

The EV charging demands depend on the EV penetration level, individual EV charging power, and EV charging scenarios. In the impact study of EV integration in this chapter, three EV penetration levels are used: 10%, 15%, and 20%. Two individual EV charging power options are considered: one-phase 16 A charging and three-phase 16 A charging. The load cases considered with different EV penetration levels are listed in Table 11.1. The five charging scenarios used in the impact study are described in Table 11.2.

11.2.4 EV Distribution Over the Grid

In the impact study, it is assumed that the EVs are distributed according to the household number in specific areas. Therefore, for the impact study of EV integration on 60 kV, 10 kV and 0.4 kV grids, the EV number is determined according to the household customer number

Table 11.1 Overview of load cases considered.

Load case	No. of EVs	No. of phases	Current (A)	Maximum charging power (kW)	EV penetration (%)
1	2000	1	16	3.7	10
2	2000	3	16	11.0	10
3	3000	1	16	3.7	15
4	3000	3	16	11.0	15
5	4000	1	16	3.7	20
6	4000	3	16	11.0	20

connected to 60/10 kV primary substations at the 60 kV grid, the household customer number connected to secondary substations at the 10 kV grid, and the household number connected at the 0.4 kV busbars at the 0.4 kV grid.

11.2.5   Loading Limits

In order to ensure that there is enough capacity in the grid to handle both normal and contingency conditions, the N-1 study should be done. However, in order to finish the EV impact study in a realistic time frame, a compromise has been made according to the DSO practice.

For the 60 kV grid, the loading limits for lines in the ring are set at 50%; 100% is used for other lines. The loading limits for transformers are 100%.

The loading limits for 10 kV cables have been set as 67%. The assumption is that two neighboring feeders will be used to resupply a feeder with a fault, and the two feeders have the same total capacity and will divide the load from the faulty feeder equally. The result is

Table 11.2 Definition of EV charging scenarios.

Charge strategy	Description
Dumb charge	Each EV is charging each time it is connected (i.e., not driving) up to full contractural state of charge (COSOC; i.e. 85% of full charge).
Dumb charge home	Each EV is charging up to full COSOC when it returns home after the last driving tour of the day.
Timer	Each EV starts charging at a certain time of the day (e.g., 22:00) if it is connected. If not connected it will start charging when it returns back home after 22:00 and will charge up to full COSOC if not disconnected before.
Fleet all day	Each EV will be optimally charged up to full COSOC each time the EV is connected (i.e., not driving) also during the day when the cost of electricity is high.
Fleet night	Each EV will be optimally charge up to full COSOC each day in the time period when the electricity cost is lowest. This is the optimal strategy (also from a theoretical point of view).

that the two feeders will each pick up 33% of the load from the faulty feeder, giving a load for the two feeders of 100%.

For the 0.4 kV grid, the loading limit is set as 100%.

11.2.6   Limitations

The geographic information has not been included in the grid model. Therefore, the roaming effect of EVs has not been handled in a precise way.

The study results are quite dependent on driving data. The widely distributed dumb charging demands might need to be updated with real EV driving data or more accurate driving data. In the existing driving data, the driving pattern of weekdays is not differentiated. However, they are quite different in reality, especially on Fridays.

The loading limits may differ depending on the location, components (overhead lines or cables), and operation guidelines of the grid. Therefore, the results from the Edison project should not be used as a general conclusion for the EV grid impact study.

11.3   Bornholm Power System

The Bornholm power system is a Danish electricity distribution system of the Bornholm island, which is situated just south of Sweden. Østkraft is the DSO supplying electricity to more than 28 000 customers on Bornholm. The peak load was 56 MW in 2007. The Bornholm power system is part of the Nordic interconnected power system and power market, and it has many of the characteristics of a typical Danish distribution system. With respect to area, electricity demand, and population, Bornholm corresponds to approximate 1% of Denmark. The wind power penetration in 2007 was more than 30%, and the system can be operated isolated in an island mode. The Bornholm power system, therefore, is a unique facility for experiments with new smart grid technologies.

11.3.1   Overview of Bornholm Power System

The main features of the Bornholm power system which make it a perfect test site for smart grid technologies are:

- 33% wind power penetration;
- islanding operation capability;
- strong political support and public understanding for using renewable energy;
- green energy strategy – 100% RES-based island "'Bright Green Island";
- the DSO, Østkraft, has joined the consortium of PowerLabDK.

The main components of the Bornholm power system are a 132/60 kV substation in Sweden, a connection between Sweden and Bornholm, 60 kV network, 10 kV network, 0.4 kV network, load, customers, generation units, control room, communication system, biogas plant "Biokraft", and district heating systems.

In the Borrby 132/60 kV substation, there are two 132/60 kV transformers. One of them is connected to the Bornholm power system. The connection from the Bornholm power system to Sweden comprises overhead lines and cables.

The 60 kV Bornholm power system consists of 16 60/10 kV substations and 60 kV cables and overhead lines connecting the substations.

The 10 kV Bornholm power system is comprised of 91 feeders which are made up of 184 km of overhead lines and 730 km of cables.

The 0.4 kV Bornholm power system consists of 10/0.4 kV substations, overhead lines, and cables. The details of the 0.4 kV Bornholm power system are:

- 1006 10/0.4 kV substations;
- 478 km of overhead lines;
- 1409 km of cables.

The power generators on Bornholm include 1 big thermal power plant (Block 5), 1 big combined heat and power plant, 14 diesel generators, 2 biomass generators, and 30 wind turbine generators. The total generation capacity is 133.5 MW.

The loads of the Bornholm power system in 2007 are peak load 56 MW and energy consumption 268 GWh.

11.3.2 Bornholm Power System Model in PowerFactory

The Bornholm power system in DigSILENT's PowerFactory has been used in order to analyze the impact of different EV charging scenarios with different EV penetration levels on electric distribution networks. The three parts of the Bornholm grid model in PowerFactory are presented in the following three sections.

The 60 kV Grid Model in PowerFactory

The single line diagram (SLD) of the Bornholm power system is shown in Figure 11.2. In the Bornholm power system, there are 16 60/10 kV primary substations for the 60 kV grid. The 60/10 kV primary substations are connected by 60 kV overhead lines and cables, which makes a meshed grid. The 60 kV Bornholm grid is connected to the Swedish system by sea cables and overhead lines. The Hasle primary substation is the connection point of the Bornholm power system to the sea cables, which is shown in the SLD of the Bornholm power system.

The Bornholm 60 kV grid model in PowerFactory is shown in Figure 11.3. The model is comprised of the topology of the 60 kV grid and includes cables, overhead lines and transformers, connection to the Swedish power system, including overhead lines, cables, and 135/65 kV transformer, and lump demands at the 10 kV side of each primary substation.

The 10 kV Grid in PowerFactory

The 10 kV grid model in PowerFactory consists of all the 10/0.4 kV secondary substations (1007 secondary substations), lumped load in each secondary substation, all the 10 kV line

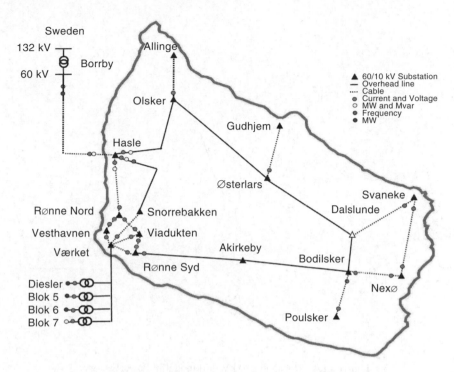

Figure 11.2   SLD of Bornholm 60 kV system.

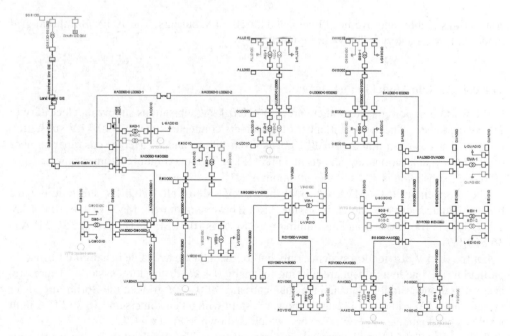

Figure 11.3   Bornholm 60 kV grid model in PowerFactory

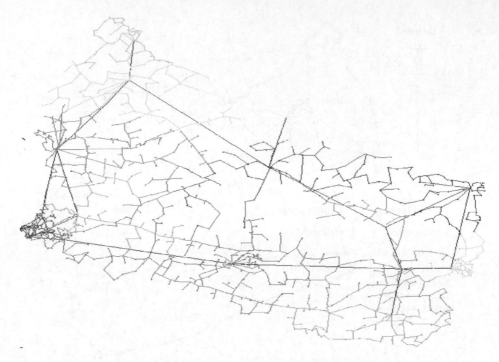

Figure 11.4   Bornholm 10 kV grid model in PowerFactory

connections (cables and overhead lines), and all 10.5 kV busbars. The 10 kV Bornholm grid model in PowerFactory is shown in Figure 11.4.

The 0.4 kV Grid Model in PowerFactory

In the Bornholm power system, there are 1006 10/0.4 kV substations supplying electricity to 28 289 customers. The average number of customers connected to each 10/0.4 kV substation is 28.09. The customer numbers of all 10/0.4 kV substations and the average customer number of 10/0.4 kV substations are shown in Figure 11.5. The maximum customer number of a 10/0.4 kV substation is 323 and the minimum is 1.

The short circuit (SC) MVA values at the 10 kV side of all 10/0.4 kV substations have been calculated and are shown in Figure 11.6. The maximum SC MVA of the 10/0.4 kV substations is 100.15 MVA and the minimum SC MVA is 10.01 MVA. The average SC MVA is 36.66 MVA.

For the 0.4 kV distribution systems, only half of the data are available in a digital format. In particular, the line length data are missing for many 0.4 kV distribution systems. Therefore, according to the available data and the characteristics of 0.4 kV grids of the Bornholm power system, a 0.4 kV grid of the Bornholm power system with 64 customers and 46.89 MVA fault level has been modeled in PowerFactory, which is shown in Figure 11.7.

In the selected 0.4 kV grid, there are six customer types. The details of the customer number and types are listed in Table 11.3.

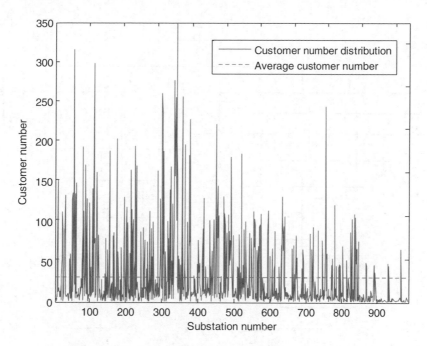

Figure 11.5 Customer number per substation on Bornholm

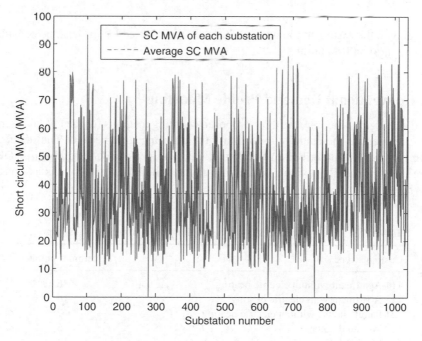

Figure 11.6 SC MVA per substation on Bornholm.

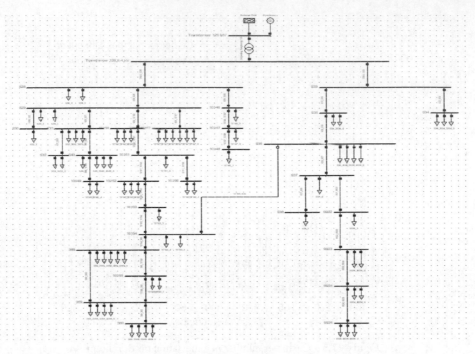

Figure 11.7   Bornholm 0.4 kV grid model in PowerFactory.

According to the customer number and SC level, the selected LV grid can be considered as a typical LV grid on Bornholm.

11.4   Conventional Demand Profile Modeling

For the EV charging schedule, the three proposed charging methods for the EV home charging are used to calculate the charging power and the charging time needed. The details of the charging methods are described in Section 11.2.3. The personal car numbers in DK1 and DK2 are used for the EV charging schedule study and the EV battery size is assumed to be 23.3 kWh.

Table 11.3   Customer type and number of the typical Bornholm 0.4 kV grid.

Customer type	Customer number
111 Apartment without electric heating	36
112 Apartment with electric heating	14
121 Family house without electric heating	7
130 Weekend cottage	5
441 Electricity, gas, water, and heat supply	1
446 Postal services and telecommunication	1

In the Bornholm power system modeling work, another important aspect is to model the end-user demand profile as realistically as possible. From the Bornholm power system database, the demand data of end-users are yearly consumption. Therefore, it is a challenge to get realistic demand profiles.

In order to obtain realistic demand profiles, the averaged demand profiles of different customer types from Elforbrugspanel (http://www.elforbrugspanel.dk) are used to determine the demand profile of customers in the Bornholm power system based on the yearly electricity consumption. The average electricity consumptions of different customer types [10] are used to determine the customer types according to the end-user yearly electricity consumption.

For the impact studies of EV grid integration, time-series power-flow studies were carried out to investigate the effects of EV demands on the power system component loading and voltage profiles. Therefore, both the peak demand and the demands adjacent to peak demand are of interest in the studies. Hence, it is important to select the demand profile considering both peak demand and the adjacent demands.

In order to illustrate which demand profile should be selected, the whole-year daily demand profiles of apartments without electric heating were obtained and are shown in Figure 11.8. Besides the whole-year daily demand profiles, the daily demand profiles of each day within a week with the highest peak demand were also obtained and are shown in Figure 11.9.

It can be seen from Figure 11.9 that the demand of the day with highest peak demand can encompass most of the demand profiles. Therefore, it is acceptable to choose the demand profile of the day with highest peak demand to carry out grid impact studies of EV integration.

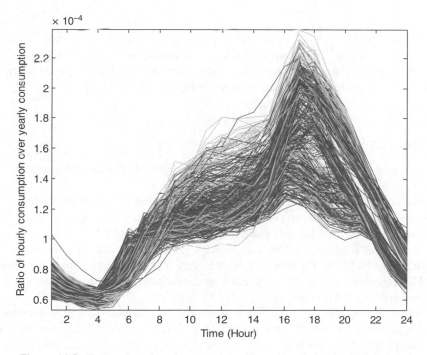

Figure 11.8 Demand profile of apartments without electric heating for all days.

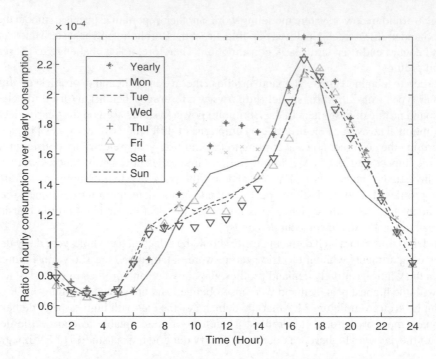

Figure 11.9 Demand profile of apartments without electric heating.

The demand profiles of other customer types were also obtained in order to carry out the grid impact studies. The demand profiles of a family house with and without electric heating, weekend cottage, electricity, gas, water, and heat supply, and postal services and telecommunication are obtained and are shown in Figures 11.10–11.14.

11.5 Impact Study on 0.4 kV Grid

The 0.4 kV grid described in Section 11.3.2 was used to carry out the EV grid impact study on 0.4 kV grids.

The results of 10% EV penetration and three-phase charging are used to illustrate the loading of transformers and cables, as shown in Figure 11.15 and Figure 11.16, respectively. It is shown that the loading of transformer(s) is much higher than that of cables.

The transformer loadings for 15% and 20% EV penetrations with one-phase charging are shown in Figure 11.17 and Figure 11.18, respectively. The maximum loading is less than the selected loading limit (100%) with 15% EV penetration and one-phase charging and 20% EV penetration and one-phase charging.

Another interesting finding is the "dumb charging" EV demands are quite distributed and the peak load with economically efficient charging scenario may be higher than the one with "dumb charging." This is due to the distribution of the time when EVs reach home, as shown in Figure 11.19.

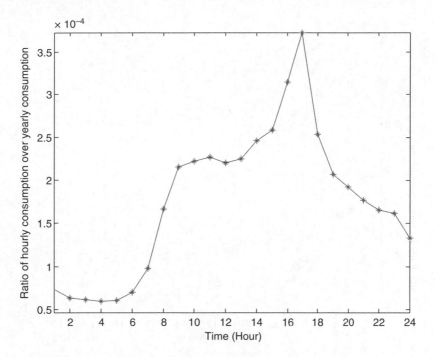

Figure 11.10 Demand profile of family house without electric heating.

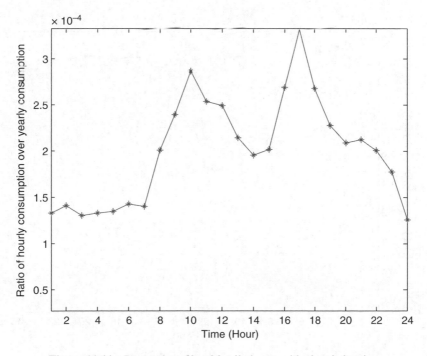

Figure 11.11 Demand profile of family house with electric heating.

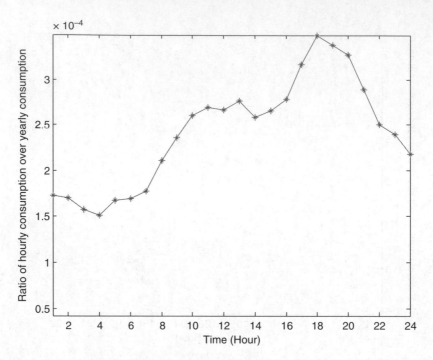

Figure 11.12   Demand profile of weekend cottages.

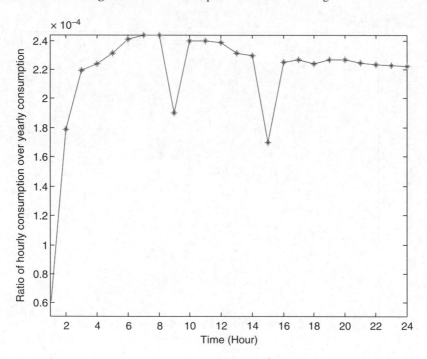

Figure 11.13   Demand profile of electricity, gas, water, and heat supply.

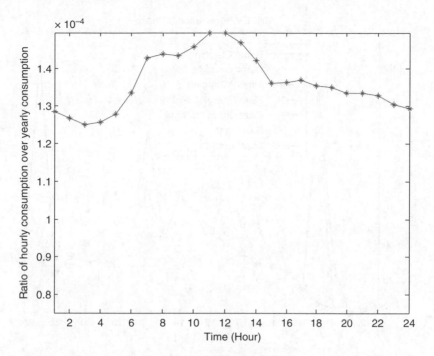

Figure 11.14 Demand profile of postal services and telecommunication.

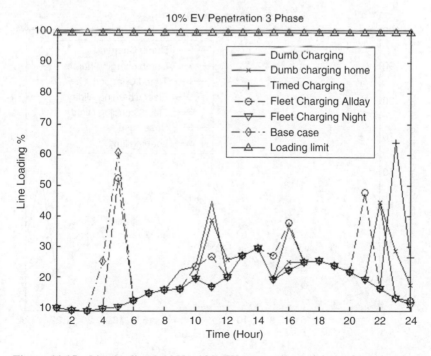

Figure 11.15 Line loading 0.4 kV – 10% EV penetration and three-phase charging.

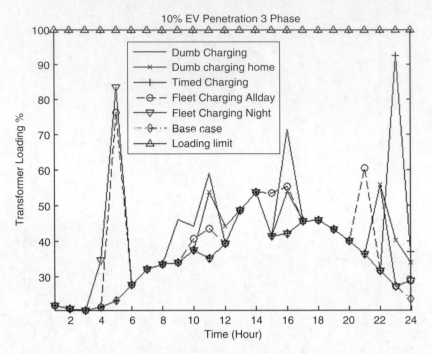

Figure 11.16   Transformer loading 0.4 kV – 10% EV penetration and three-phase charging.

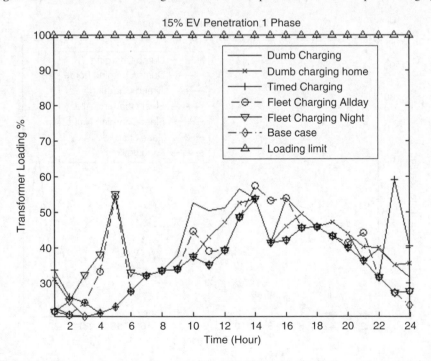

Figure 11.17   Transformer loading 0.4 kV – 15% EV penetration and one-phase charging.

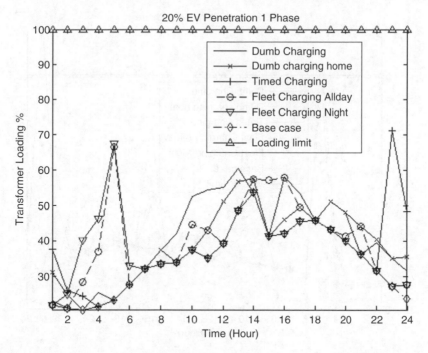

Figure 11.18 Transformer loading 0.4 kV – 20% EV penetration and one-phase charging.

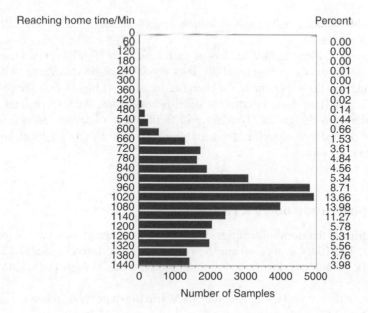

Figure 11.19 Distribution of the time when EVs reach home.

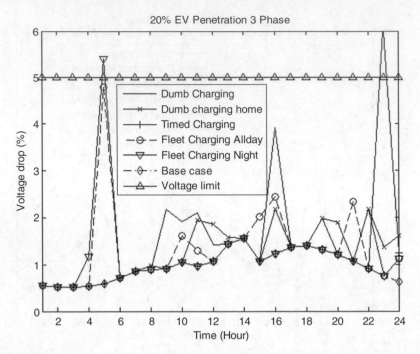

Figure 11.20 Voltage drop 0.4 kV – 20% EV penetration and three-phase charging.

However, this result has to be checked with more accurate driving data; that is, the weekdays should be differentiated.

The voltage drop along 0.4 kV feeders of each hour with 20% EV penetration and three-phase charging is shown in Figure 11.20. It is shown that the maximum voltage drop of timed charging and fleet-operator-based charging scenarios is higher than the 5% limit specified by standards. Therefore, besides the overloading issues, the undervoltage issue needs to be considered to design the dynamic grid tariffs. The undervoltage can be alleviated by more distributed EV charging demands or other measures (e.g., local compensation, bigger cables).

11.6 Impact Study on 10 kV Grid

Two 10 kV grids of the Bornholm power system were used to carry out the EV impact studies on MV grids. The 10 kV grids used are shown in Figure 11.21 and Figure 11.22. The Ronne Syd is an MV grid in the downtown area and the Svaneke MV grid is an MV grid in the rural area.

The results of the EV impact study of the 10 kV grid illustrate more or less the same findings as those of the 0.4 kV grid.

The transformers have higher loading compared with lines, which is shown in Figure 11.23 and Figure 11.24.

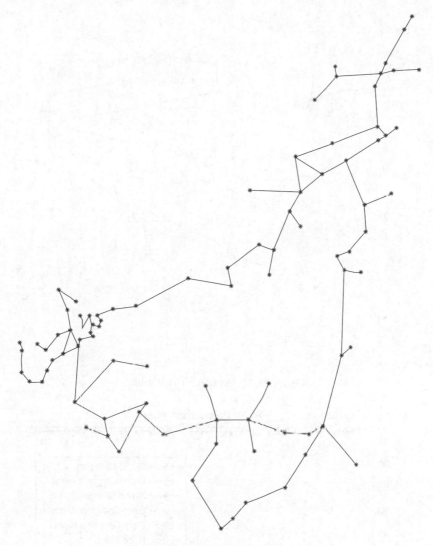

Figure 11.21   RonneSyd 10 kV grid.

The 10 kV grid can accommodate 20% EV penetration with one-phase charging without overloading issues. The transformer loading of 20% EV penetration and one-phase charging is used to illustrate the finding, which is shown in Figure 11.25.

The impact of three-phase charging is shown in Figure 11.24, which shows that the transformer loading is higher than the selected loading limit for all charging scenarios.

It is also shown that the "dumb charging" EV demands are more distributed than the "fleet charging" EV demands. This can be explained by recourse to Figure 11.19.

The maximum voltage drop along feeders of the 20% EV penetration and three-phase charging study case shown in Figure 11.26 shows that there is no voltage issue.

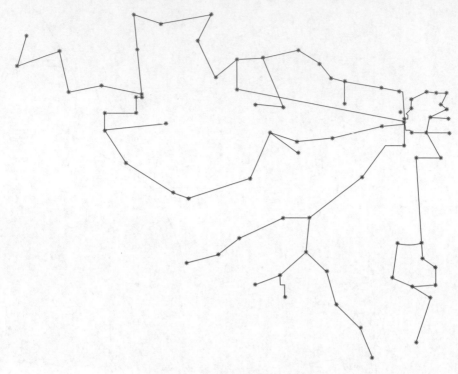

Figure 11.22   Svaneke 10 kV grid.

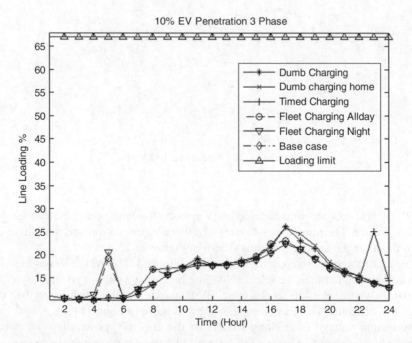

Figure 11.23   Line loading 10 kV Ronne Syd – 10% EV penetration and three-phase charging.

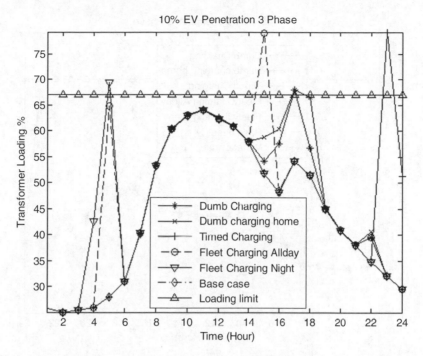

Figure 11.24 Transformer loading 10 kV Ronne Syd – 10% EV penetration and three-phase charging.

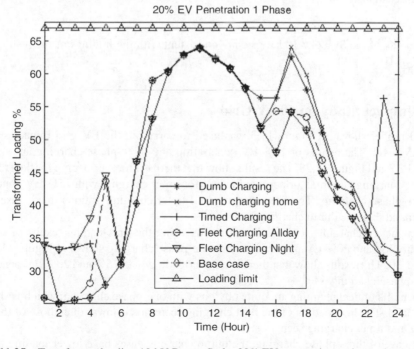

Figure 11.25 Transformer loading 10 kV Ronne Syd – 20% EV penetration and one-phase charging.

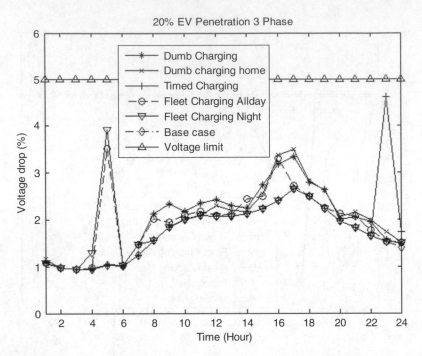

Figure 11.26 Maximum voltage drop along feeders 10 kV Ronne Syd – 20% EV penetration and three-phase charging.

The results of the Svaneke 10 kV grid are consistent with the findings of the results of the Ronne Syd 10 kV grid.

11.7 Impact Study on 60 kV Grid

The grid model shown in Figure 11.3 was used to carry out the EV grid impact study on the 60 kV level. The results of 15% EV penetration and three-phase charging are shown in Figure 11.27 and Figure 11.28. The results show that there is line overloading for fleet charging and timer charging and transformer overloading for timer charging with 15% EV penetration and three-phase charging. The fleet charging and timer charging options may cause higher peak demand than the dumb charging option.

The impact of charging power is illustrated by using the line loading results of the 15% EV penetration and 20% EV penetration with one-phase charging shown in Figure 11.29 and Figure 11.30. The results show that there is no overloading with 15% and 20% EV penetration and one-phase charging.

The benefits of fleet charging are quite obvious with one-phase charging, which is shown in Figure 11.30. The line loading of the fleet charging scenarios is lower than those of the dumb charging and timer charging scenarios.

In the case of three-phase charging, the dumb charging cases have lower loading than the fleet charging scenarios.

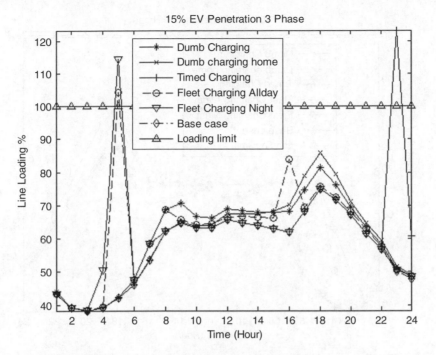

Figure 11.27   Line loading – 15% EV penetration and three-phase charging.

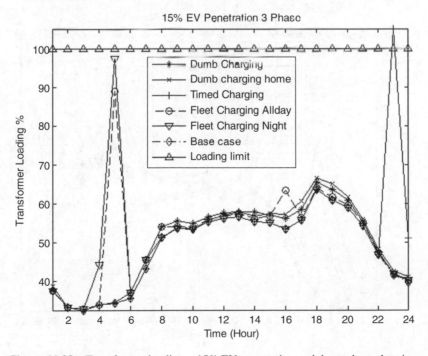

Figure 11.28   Transformer loading – 15% EV penetration and three-phase charging.

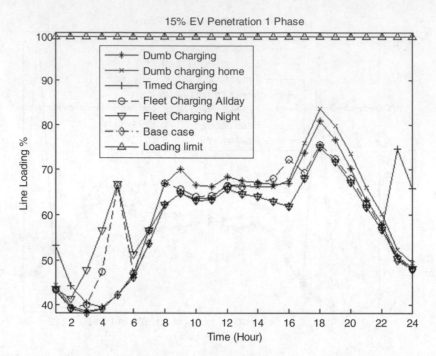

Figure 11.29   Line loading – 15% EV penetration and one-phase charging.

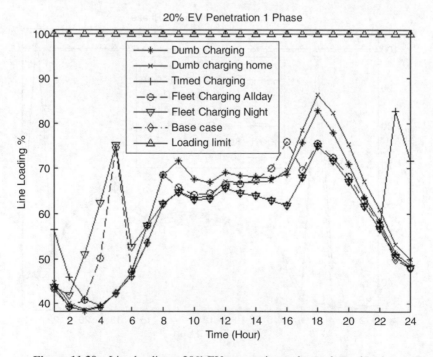

Figure 11.30   Line loading – 20% EV penetration and one-phase charging.

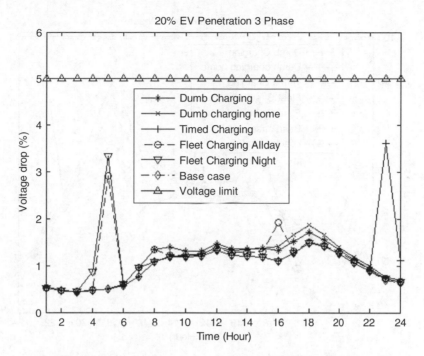

Figure 11.31   Maximum voltage drop on 60 kV grid – 20% EV penetration and three-phase charging.

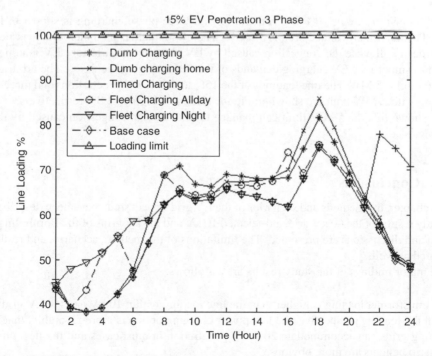

Figure 11.32   Line loading – 15% EV penetration and three-phase charging with 5 MW limit.

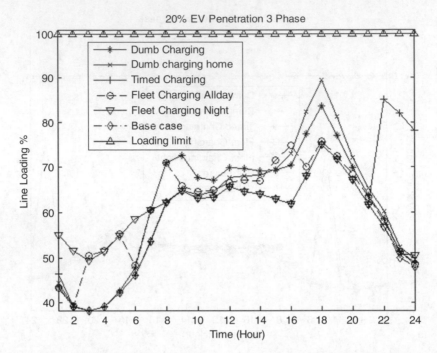

Figure 11.33   Line loading – 20% EV penetration and three-phase charging without 5 MW limit.

The voltage profile of 20% EV penetration and three-phase charging is shown in Figure 11.31. The maximum voltage drop is 3.603%. Therefore, there is no undervoltage issue.

In order to alleviate the congestion caused by EV charging, an arbitrary EV loading limit is used to smooth the EV charging demands of the fleet charging scenarios. The arbitrary EV loading limit is 5 MW. The line loadings of the 15% and 20% EV penetrations and three-phase charging with 5 MW limit are shown in Figure 11.32 and Figure 11.33, respectively.

It is shown that the 5 MW limit can manage to alleviate the congestion caused by the EV charging.

11.8   Conclusions

In this chapter, the methods and scenarios of the EV grid impact study have been described and the study results of the 60 kV grid and selected 10 kV and 0.4 kV grids of the Bornholm power system and discussions are presented. The limitations of the method, scenario, and results are presented as well.

The major findings of the study results are as follows:

- The transformer loading is higher than the line loading within 10 kV and 0.4 kV grids.
- From the loading perspective, the low power charging option is more favorable; that is, the existing grids can accommodate 20% EV penetration in most cases and the fleet charging scenario benefits are more obvious.

- There is an undervoltage issue with 20% EV penetration and three-phase 16 A charging in the 0.4 kV grid with ±5% voltage drop limit.
- The dumb charging EV demands are quite distributed owing to the quite spread distribution of the EV reaching-home time. However, this should be updated with more accurate driving data.
- In the case of three-phase 16 A charging, it is very important to design an efficient scheme to alleviate the congestion caused by EV charging demands, which is demonstrated by the impact study results on 60 kV grid with 5 MW arbitrary EV demand limit (e.g., efficient DSO market to stimulate the EV charging demand shifting, direct control).

References

[1] Lopes, J.A.P., Soares, F.J., and Almedia, P.M.R. (2009) Identifying management procedures to deal with connection of electric vehicles in the grid, in *2009 IEEE Bucharest PowerTech*, IEEE, Piscataway, NJ, pp. 1–8.

[2] Clement, K., Haesen, E., and Driesen, J. (2009) Coordinated charging of multiple plug-in hybrid electric vehicles in residential distribution grids, in *IEEE/PES Power Systems Conference and Exposition, 2009. PSCE '09*, IEEE, Piscataway, NJ, pp. 1–7.

[3] Maitra, A., Kook, K.S., Taylor, J., and Giumento, A. (2010) Grid impacts of plug-in electric vehicles on Hydro Quebec's distribution system, in *2010 IEEE PES Transmission and Distribution Conference and Exposition*, IEEE, Piscataway, NJ, pp. 1–7.

[4] Taylor, J., Maitra, M., Alexander, D., and Duvall, M. (2009) Evaluation of the impact of plug-in electric vehicle loading on distribution system operations, in *IEEE Power & Energy Society General Meeting, 2009. PES '09*, IEEE, Piscataway, NJ, pp. 1–6.

[5] Sundström, O. and Binding, C. (2010) Planning electric-drive vehicle charging under constrained grid conditions, in *2010 International Conference on Power System Technology (POWERCON)*, IEEE, Piscataway, NJ, pp. 1–6.

[6] Rotering, N. and Ilic, M. (2010) Optimal charge control of plug-in hybrid electric vehicles in deregulated electricity markets. *IEEE Transactions on Power Systems*, **26** (3), 1021–1029.

[7] Babaeim, S., Steen, D., Tuan, L.A. *et al.* (2010) Effects of plug-in electric vehicles on distribution systems: a real case of Gothenburg, in *2010 IEEE PES Innovative Smart Grid Technologies Conference Europe (ISGT Europe)*, IEEE, Piscataway, NJ, pp. 1–8.

[8] Dyke, K.J., Schofield, N., and Barnes, M. (2010) The impact of transport electrification on electrical networks. *IEEE Transactions on Industrial Electronics*, **57** (12), 3917–3926.

[9] Lopes, J.A.P., Soares, F.J., and Almedia, P.M.R. (2011) Integration of electric vehicles in the electric power system. *Proceedings of the IEEE*, **99** (1), 168–183.

[10] Dansk Energi (2010) Danish Electricity Supply 09 Statistical Survey, http://www.danishenergyassociation.com/~/media/Energi_i_tal/Statistik_09_UK.ppt.ashx (accessed January 2013).

Index

Grid Integration of Electric Vehicles in Open Electricity Markets, First Edition. Edited by Qiuwei Wu.
© 2013 John Wiley & Sons, Ltd. Published 2013 by John Wiley & Sons, Ltd.